Automatic Assembly

MANUFACTURING ENGINEERING AND MATERIALS PROCESSING

A Series of Reference Books and Textbooks

SERIES EDITORS

Geoffrey Boothroyd

Department of Mechanical Engineering
University of Massachusetts
Amherst, Massachusetts

George E. Dieter

Dean, College of Engineering
University of Maryland
College Park, Maryland

1. Computers in Manufacturing, *U. Rembold, M. Seth, and J. S. Weinstein*
2. Cold Rolling of Steel, *William L. Roberts*
3. Strengthening of Ceramics: Treatments, Tests, and Design Applications, *Henry P. Kirchner*
4. Metal Forming: The Application of Limit Analysis, *Betzalel Avitzur*
5. Improving Productivity by Classification, Coding, and Data Base Standardization: The Key to Maximizing CAD/CAM and Group Technology, *William F. Hyde*
6. Automatic Assembly, *Geoffrey Boothroyd, Corrado Poli, and Laurence E. Murch*

OTHER VOLUMES IN PREPARATION

Automatic Assembly

GEOFFREY BOOTHROYD

CORRADO POLI

LAURENCE E. MURCH

Department of Mechanical Engineering
University of Massachusetts at Amherst
Amherst, Massachusetts

MARCEL DEKKER, INC.   New York and Basel

LIBRARY OF CONGRESS CATALOGING IN PUBLICATION DATA

Boothroyd, G. (Geoffrey), [date]
   Automatic assembly.

   (Manufacturing engineering and materials pro-
cessing; 6)
   1. Assembling machines. I. Poli, C., [date]
II. Murch, Laurence E., [date] III. Title.
IV. Series.
TJ1317.B66 1982   670.42'7   81-15180
ISBN 0-8247-1531-4   AACR2

The figures and tables listed below were previously published in
G. Boothroyd and A. H. Redford, *Mechanized Assembly*, McGraw-Hill,
London, 1968, which is now out of print. They are reproduced here
courtesy of the authors. Figures 2.1 - 2.15, 3.1 - 3.8, 3.13, 3.14,
3.16, 3.17, 4.2, 4.6, 4.7, 4.8, 4.10, 4.11, 4.15, 4.17 - 4.19, 4.22,
4.23, 4.48, 4.54, 5.2, 5.4 - 5.7, 5.10 5.44 - 5.46, 6.1, 6.8 - 6.11,
6.24 - 6.40, 7.1, 7.3 - 7.5, 7.7, 8.1 - 8.8, 8.10, 8.11, 9.1 - 9.9,
II.1, II.2, IV.1 - IV.6, and Tables 9.1 - 9.4.

MARCEL DEKKER, INC.
270 Madison Avenue, New York, New York 10016

Current printing (last digit):
10 9 8 7 6 5 4 3 2 1

PRINTED IN THE UNITED STATES OF AMERICA

Preface

The subjects of this book—parts feeding and orientation and other aspects of automatic assembly, including its economics—are rapidly growing in importance as industry seeks ways to improve manufacturing productivity. There is a strong desire in many companies for manufacturing engineers and designers to learn about automatic assembly and to exploit the various techniques being developed.

Unfortunately, automatic assembly is popularly thought of more as an art than as a science. This view tends to deter manufacturing engineers from considering the application of automation to assembly. This book is directed mainly at manufacturing engineers, and also at mechanical engineers, in the hope that the fundamental approach employed will help to correct this situation. It is also hoped that the book will serve as a text for those studying manufacturing engineering in technical colleges and universities. For this reason, a series of problems is presented at the end of the book and Appendix IV describes two typical experiments suitable for student laboratory work.

Parts of this book were originally published in 1968 under the title *Mechanized Assembly*, by G. Boothroyd and A. H. Redford. The original material developed at the University of Salford has now been considerably updated by the addition of work carried out at the University of Massachusetts by the present authors and numerous students, and supported by several grants from the National Science Foundation.

The present authors wish to thank Dr. A. H. Redford for his kind permission to use material published in the original book and the National Science Foundation for their support. They also wish to thank Wendy Fortin for preparation of the artwork and Janet Boothroyd for typing and editing the manuscript.

Geoffrey Boothroyd
Corrado Poli
Laurence E. Murch

Contents

PREFACE   *iii*

Chapter 1
INTRODUCTION   *1*

1.1 Historical Development of the Assembly Process   *2*
1.2 Choice of Assembly Method   *5*
1.3 Advantages of Automatic Assembly   *5*

Chapter 2
TRANSFER SYSTEMS   *9*

2.1 Continuous Transfer   *9*
2.2 Intermittent Transfer   *12*
2.3 Indexing Mechanisms   *19*
2.4 Operator-Paced Free-Transfer Machine   *25*

Chapter 3
VIBRATORY BOWL FEEDERS   *27*

3.1 Mechanics of Vibratory Conveying   *27*
3.2 Effect of Frequency   *32*
3.3 Effect of Track Acceleration   *33*
3.4 Effect of Vibration Angle   *34*
3.5 Effect of Track Angle   *35*

3.6   Effect of Coefficient of Friction   37
3.7   Estimating the Mean Conveying Velocity   37
3.8   Load Sensitivity   41
3.9   Effect of Load Sensitivity on Recirculation of Parts   43
3.10 Solutions to Load Sensitivity   46
3.11 Orientation of Parts   48
3.12 Bowl Feeder Design   48
3.13 Spiral Elevators   49

Chapter 4
NONVIBRATORY FEEDERS   51

4.1   Reciprocating Tube Hopper Feeder   52
4.2   Centerboard Hopper Feeder   56
4.3   Reciprocating Fork Hopper Feeder   64
4.4   External Gate Hopper Feeder   65
4.5   Rotary Disk Feeder   70
4.6   Centrifugal Hopper Feeder   76
4.7   Revolving Hook Hopper Feeder   80
4.8   Stationary Hook Hopper Feeder   82
4.9   Bladed Wheel Hopper Feeder   86
4.10 Tumbling Barrel Hopper Feeder   89
4.11 Rotary Centerboard Hopper Feeder   93
4.12 Magnetic Disk Feeder   94
4.13 Elevating Hopper Feeder   96
4.14 Magnetic Elevating Hopper Feeder   99
4.15 Magazines   99

Chapter 5
ORIENTATION OF PARTS   101

5.1 A Typical Orienting System   101
5.2 Effect of Active Orienting Devices on Feed Rate   107
5.3 Analysis of Orienting Systems   108
5.4 Performance of an Orienting Device   117
5.5 Natural Resting Aspects of Parts for
     Automatic Handling   126
5.6 Analysis of a Typical Orienting System   137
5.7 Out-of-Bowl Tooling   145

Chapter 6
FEED TRACKS, ESCAPEMENTS, PARTS-PLACING
MECHANISMS, AND ROBOTS   149

6.1 Gravity Feed Track Arrangements   149
6.2 Powered Feed Tracks   180
6.3 Escapements   184
6.4 Parts-Placing Mechanisms   191
6.5 Assembly Robots   195

Chapter 7
PERFORMANCE AND ECONOMICS OF ASSEMBLY SYSTEMS   201

7.1   Indexing Machines   202
7.2   Free-Transfer Machines   209
7.3   Basis of Economic Comparisons for Automation
      Equipment   218
7.4   Comparison of the Economics of Free-Transfer
      and Indexing Machines   220
7.5   Multistation Hybrid System (Indexing Machines
      Linked by Buffer Storage)   223
7.6   Multistation Operator Assembly   225
7.7   Single-Station Operator Assembly   227
7.8   Programmable Assembly Automation   228
7.9   Single-Station Assembly Center   234
7.10 Comparison of Different Assembly Systems   240

Chapter 8
DESIGN FOR AUTOMATIC ASSEMBLY   255

8.1 Product Design for Ease of Assembly   256
8.2 Design of Parts for Feeding and Orienting   262
8.3 Summary of Design Rules   266
8.4 Analysis of Design for Assembly   267

Chapter 9
DESIGN OF ASSEMBLY MACHINES   275

9.1 Design Factors to Reduce Machine Downtime
     Due to Defective Parts   276
9.2 Feasibility Study   279

PROBLEMS   *299*

Appendix I
SIMPLE METHOD FOR THE DETERMINATION OF
THE COEFFICIENT OF DYNAMIC FRICTION   *313*

I.1 The Method   *313*
I.2 Analysis   *315*
I.3 Precision of the Method   *318*
I.4 Discussion   *318*

Appendix II
OUT-OF-PHASE VIBRATORY CONVEYORS   *319*

II.1 Out-of-Phase Conveying   *320*
II.2 Practical Applications   *322*

Appendix III
U. MASS. CODING SYSTEM FOR THE AUTOMATIC
FEEDING AND ORIENTING OF SMALL PARTS   *325*

III.1 Introduction   *325*
III.2 Terminology   *325*
III.3 Examples   *333*
III.4 Coding System for Small Parts for Automatic
       Handling (Choice of the First Digit)   *335*

Appendix IV
LABORATORY EXPERIMENTS   *355*

IV.1 Performance of a Vibratory Bowl Feeder   *359*
IV.2 Performance of a Horizontal Delivery Gravity Feed Track *359*

NOMENCLATURE   *365*

INDEX   *373*

Automatic Assembly

Chapter 1

Introduction

The increasing need for finished goods in large quantities has, in the past, led engineers to search for and to develop new methods of production. Many individual developments in the various branches of manufacturing technology have been made and have allowed the increased production of improved finished goods at lower cost. One of the most important manufacturing processes is the assembly process. This process is required when two or more component parts are to be brought together to produce the finished product.

The early history of assembly process development is closely related to the history of the development of mass-production methods. Thus, the pioneers of mass production are also the pioneers of the modern assembly process. Their new ideas and concepts have brought significant improvements in the assembly methods employed in large-volume production.

However, although some branches of manufacturing engineering, such as metal cutting and metal forming processes, have recently been developing very rapidly, the technology of the basic assembly process has failed to keep pace. Table 1.1 shows that in the United States the percentage of the total labor force involved in the assembly process varies from about 20% for the manufacture of farm machinery to almost 60% for the manufacture of telephone and telegraph equipment. Because of this, assembly costs often account for more than 50% of the total manufacturing costs. Statistical surveys show that these figures are increasing every year.

In the past few years, certain efforts have been made to reduce assembly costs by the application of automation and modern techniques, such as ultrasonic welding and die-casting. However, success has been very limited and many assembly operators are still using the same basic tools as those employed at the time of the Industrial Revolution.

Table 1.1 Percentage of Production Workers Involved in Assembly

Industry	Percentage of workers involved in assembly
Motor vehicles	45.6
Aircraft	25.6
Telephone and telegraph	58.9
Farm machinery	20.1
Household refrigerators and freezers	32.0
Typewriters	35.9
Household cooking equipment	38.1
Motorcycles, bicycles, and parts	26.3

Source: 1967 Census of Manufactures, U.S. Bureau of the Census.

1.1 Historical Development of the Assembly Process

In the early days of manufacturing technology, the complete assembly of a product was carried out by a single operator and usually, this operator also manufactured the individual component parts of the assembly. Consequently, it was necessary for the operator to be an expert in all the various aspects of the work, and training a new operator was a long and expensive task. The scale of production was often limited by the availability of trained operators rather than by the demand for the product.

In 1798, the United States needed a large supply of muskets and federal arsenals could not meet the demand. Because war with the French was imminent, it was also not possible to obtain additional supplies from Europe. However, Eli Whitney, now recognized as one of the pioneers of mass production, offered to contract to make 10,000 muskets in 28 months. Although it took 10 1/2 years to complete the contract, Whitney's novel ideas on mass production had been successfully proved. The factory at New Haven, Connecticut, built specially for the manufacture of the muskets, contained machines for producing interchangeable parts. These machines reduced the skills required by the various operators and allowed significant increases in the rate of production. In an historic demonstration in 1801, Whitney surprised his distinguished visitors when he assembled musket locks after randomly selecting parts from a heap.

The results of Eli Whitney's work brought about three primary developments in manufacturing methods. First, parts were manufactured on machines, resulting in a consistently higher quality than that of hand-made parts. These parts were now interchangeable and as a consequence assembly work was simplified. Second, the accuracy of the final product could be maintained at a higher standard, and third, production rates could be significantly increased.

Oliver Evans's conception of conveying materials from one place to another without manual effort led eventually to further developments in automation for assembly. In 1793, he used three types of conveyors in an automatic flour mill, which required only two operators. The first operator poured grain into a hopper and the second filled sacks with the flour produced by the mill. All the intermediate operations were carried out automatically with conveyors carrying the material from operation to operation.

The next significant contribution to the development of assembly methods was made by Elihu Root. In 1849, Elihu Root joined the company that was producing Colt "six-shooters." Even though at that time the various operations of assembling the component parts were quite simple, he divided these operations into basic units that could be completed more quickly and with less chance of error. Root's division of operations gave rise to the concept "divide the work and multiply the output." Using this principle, assembly work was reduced to very basic operations and with only short periods of operator training, high efficiencies could be obtained.

Frederick Winslow Taylor was probably the first person to introduce the methods of time and motion study to manufacturing technology. The object was to save the operator's time and energy by making sure that the work and all things associated with the work were placed in the best positions for carrying out the required tasks. Taylor also discovered that any worker has an optimum speed of working which, if exceeded, results in a reduction in overall performance.

Undoubtedly, the principal contributor to the development of production and assembly methods was Henry Ford. He described his principles of assembly in the following words:

"First, place the tools and then men in the sequence of the operations so that each part shall travel the least distance whilst in the process of finishing.

"Second, use work slides or some other form of carrier so that when a workman completes his operation he drops the part always in the same place which must always be the most convenient place to his hand and if possible have gravity carry the part to the next workman.

"Third, use sliding assembly lines by which parts to be assembled are delivered at convenient intervals, spaced to make it easier to work on them."

These principles were gradually applied in the production of the Model T Ford automobile.

The modern assembly line technique was first employed in the assembly of a flywheel magneto. In the original method, one operator assem-

bled a magneto in 20 min. It was found that when the process was
divided into 29 individual operations, carried out by separate operators
working at assembly stations spaced along an assembly line, the total
assembly time was reduced to 13 min 10 s. When the height of the
assembly line was raised by 8 in., the time was reduced to 7 min.
After further experiments were carried out to find the optimum speed
of the assembly line conveyor, the time was reduced to 5 min, which
was only one-fourth of the time taken by the original process of
assembly. This result encouraged Henry Ford to utilize his system
of assembly in other departments of the factory, which were producing
subassemblies for the car. Subsequently, this brought a continuous
and rapidly increasing flow of subassemblies to the operators working
on the main car assembly. It was found that the operators could not
cope with the increased flow, and it soon became clear that the main
assembly would also have to be carried out on an assembly line. At
first, the movement of the main assemblies was achieved simply by
pulling them by a rope from station to station. However, even this
development produced the amazing result of a reduction in the total
time of assembly from 12 h 28 min to 5 h 50 min. Eventually, a power-
driven endless conveyor was installed. It was flush with the floor
and wide enough to accommodate a chassis. Space was provided for
workers to either sit or stand while they carried out their operations
and the conveyor moved at a speed of 6 ft/min past 45 separate workstations. With the introduction of this conveyor, the total assembly
time was reduced to 93 min. Further improvements led to an even
shorter overall assembly time and eventually, a production rate of
one car every 10 s of the working day was achieved.

Although Ford's target of production had been exceeded and the
overall quality of the product had improved considerably, the assembled
products sometimes varied from the precise standards of the hand-built
prototypes. Eventually, a method of isolating difficulties and correcting
them in advance was adopted before mass production began. The method
was basically to set up a pilot plant, where a complete assembly line
was installed, using exactly the same tools, templates, forming devices,
gauges, and even the same labor skills that would eventually be used
for mass production. This method has now become standard practice
for all large assembly plants.

The type of assembly operation dealt with above is usually referred
to as operator assembly, and it is still the most widespread method of
assembling mass- or large-batch-produced products. However, in
certain cases, more refined methods of assembly have now emerged.

As a logical extension of the basic assembly line principle, methods
of replacing operators by mechanical means of assembly have been
devised. Here, it is usual to attempt to replace operators with automatic workheads where the tasks being performed are very simple and

to retain the operators for tasks that would be uneconomical to mechanize. This method of assembly has rapidly gained popularity for mass production and is usually referred to as automatic assembly. However, complete automation where the product is assembled completely by machine is essentially nonexistent.

1.2 Choice of Assembly Method

When considering the assembly of a product, a manufacturer has to take into account the many factors that affect the choice of assembly system. For a new product, the following considerations are generally important:

1. Cost of assembly
2. Production rate required
3. Availability of labor
4. Market life of the product

If an attempt is to be made to justify the automation of an existing operator assembly line, consideration has to be given to the redeployment of those operators who would become redundant. If labor is plentiful, the degree of automation depends on the reduction in cost of assembly and the increase in production rate brought about by the automation of the assembly line. However, it must be remembered that, in general, the capital investment in automatic machinery has to be amortized over the market life of the product unless the machinery may be adapted to assemble a new product. It is clear that if this is not the case and the market life of the product is short, automation is generally not justifiable.

A shortage of labor may often lead a manufacturer to consider automatic assembly when in fact it can be shown that operator assembly would be cheaper. Conversely, a manufacturer may be unable to automate because suitable employment cannot be found for the operators who would become redundant. Another reason for considering automation in a situation where operator assembly would be more economical is on a research basis, to gain experience in the field.

1.3 Advantages of Automatic Assembly

Following are some of the advantages of automation:

1. Reduction in the cost of assembly
2. Increased productivity

3. A more consistent product
4. Removal of operators from hazardous operations

A reduction in costs is often the main consideration and, except for the special circumstances listed above, it could be expected that automation would not be carried out if it was not expected to produce a reduction in costs.

Productivity in an advanced industrial society is an important measure of operating efficiency. Increased productivity, although not directly beneficial to a manufacturer unless labor is scarce, is necessary to an expanding economy because it releases personnel for other tasks. It is clear that when put into effect, automation of assembly lines generally reduces the number of operators required and hence increases productivity.

Some of the assembly tasks that an operator can perform easily are extremely difficult to duplicate on even the most sophisticated automatic workhead. An operator can often carry out a visual inspection of the part to be assembled, and parts that are obviously defective can be discarded. Sometimes a very elaborate inspection system is required to detect even the most obviously defective part. If an attempt is made to assemble a part that appears to be acceptable but is in fact defective, an operator, after unsuccessfully trying to complete the assembly, can reject the part very quickly without a significant loss in production. In automatic assembly, however, unless the part has been rejected by the feeding device, an automatic workhead will probably stop and time will then be wasted locating and eliminating the fault. If a part has only a minor defect, an operator may be able to complete the assembly, but the resulting product may not be completely satisfactory. It is often suggested that one of the advantages of automatic assembly is that it ensures a product of consistently high quality because the machine faults if the parts do not conform to the required specification.

In some situations, assembly by operators would be hazardous due to high temperatures and the presence of toxic substances and other materials. Under these circumstances, assembly by mechanical means is obviously advantageous.

An automatic assembly machine usually consists of a transfer system for moving the assemblies from workstation to workstation, automatic workheads to perform the simple assembly operations, vacant workstations for operators to carry out the more complicated assembly operations, and inspection stations to check that the various operations have been completed successfully. The automatic workheads are either fed manually with individual or magazine-stored component parts or are supplied with parts from an automatic feeder (often a vibratory bowl feeder) through a feed track. The workheads themselves usually consist of either a fastening device or a parts-placing mechanism. Examples of these workheads are nut and screw running heads, welding heads, riveting heads, soldering heads, push and guide placing mechanisms, and pick and place mechanisms.

In the following chapters the basic components of assembly machines are discussed separately and then the overall performance and design of assembly machines are discussed.

Chapter 2

Transfer Systems

In automatic assembly the various individual assembly operations are generally carried out at separate workstations. For this method of assembly, a system is required for transferring the partly completed assemblies from workstation to workstation, and a means must be provided of ensuring that no relative motion exists between the assembly and the workhead while the operation is being carried out. As the assembly passes from station to station, it is necessary that it be maintained in the required attitude. For this purpose, the assembly is usually built up on a base or *work carrier* and the machine is designed to transfer the work carrier from station to station; an example of a typical work carrier is shown in Fig. 2.1. Assembly machines are usually classified according to the system adopted for transferring the work carriers (Fig. 2.2). Thus, an in-line assembly machine is one where the work carriers are transferred *in-line* along a straight slideway, and a *rotary* machine is one where the work carriers move in a circular path. In both types of machine, the transfer of work carriers may be *continuous* or *intermittent*.

2.1 Continuous Transfer

With continuous transfer the work carriers are moving at constant speed while the workheads index back and forth (Fig. 2.3). In this case, the assembly operations are carried out during the period when the workheads are moving forward, keeping pace with the work carriers; the workheads then return quickly to their initial positions, ready to repeat the operations during the next cycle. Continuous transfer systems have limited application in automatic assembly because

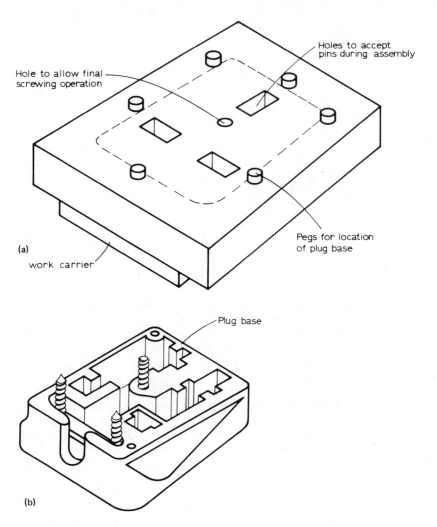

Holes to accept
pins during assembly

Hole to allow final
screwing operation

(a)

Pegs for location
of plug base

work carrier

Plug base

(b)

Fig. 2.1 Work carrier suitable for holding and transferring three-pin power plug base.

Fig. 2.2 Basic types of assembly machines.

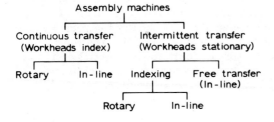

Assembly machines

Continuous transfer
(Workheads index)

Intermittent transfer
(Workheads stationary)

Rotary   In-line   Indexing   Free transfer
(In-line)

Rotary   In-line

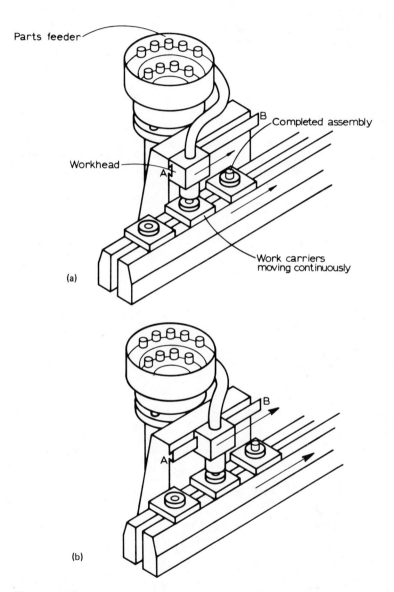

Parts feeder

B Completed assembly

Workhead

A

Work carriers
moving continuously

(a)

B

A

(b)

Fig. 2.3 In-line continuous transfer machine: (a) operation starts
with workhead at A; (b) workhead moves from A to B, keeping pace
with the workcarrier.

the workheads and associated equipment are often heavy and therefore difficult to accelerate and decelerate at the rates required. It is also often difficult under these circumstances to maintain sufficiently accurate alignment between the workheads and work carriers during the operation cycle.

2.2 Intermittent Transfer

Intermittent transfer is the more common system employed for both rotary and in-line machines. As the name implies, the work carriers are transferred intermittently and the workheads remain stationary. Usually, the transfer of all work carriers occurs simultaneously and the carriers then remain stationary to allow time for the assembly

Fig. 2.4 Rotary indexing machine.

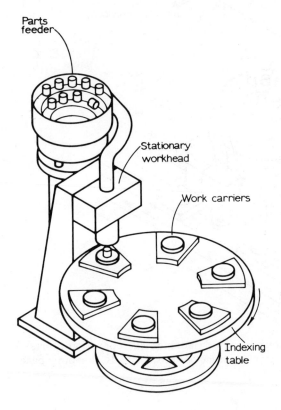

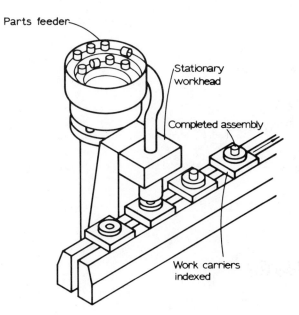

Fig. 2.5 In-line indexing machine.

operations. These machines may be termed *indexing* machines, and typical examples of the rotary and in-line types of indexing machines are shown in Figs. 2.4 and 2.5, respectively. With the rotary indexing machine, indexing of the table brings the work carriers under the various workheads in turn and assembly of the product is completed during one revolution of the table. Thus, at the appropriate station, a completed product may be taken from the machine after each index. The in-line indexing machine works on a similar principle, but in this case a completed product is removed from the end of the line after each index. With in-line machines, provision must be made for returning the empty work carriers to the beginning of the line. The transfer mechanism on in-line machines is generally one of three types: the walking beam, the shunting work carrier, or the chain-driven work carrier.

The various stages in the operation of the walking beam are illustrated in Fig. 2.6. The mechanism consists of a fixed rail provided with grooves for location of the work carriers. A transfer rail is driven in such a way that it periodically picks up a series of work carriers and deposits them farther along the fixed rail. To accomplish this, the transfer rail is attached by a linkage to a slider which is

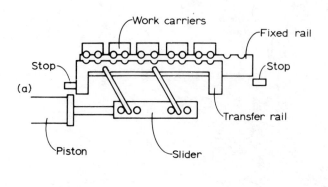

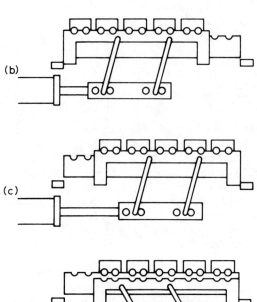

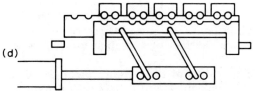

Fig. 2.6 Walking beam transfer system.

constrained to move horizontally and is activated by a piston. Figure 2.6a shows the start of the cycle where the work carriers are resting on the fixed rail and are awaiting the next index. In Fig. 2.6b it can be seen that as the slider moves to the left, the transfer rail lifts the work carriers from the fixed rail. At this point the supporting linkage has moved just past the vertical and is held in position by a stop. The piston now forces the slider to the right and the transfer rail and work carriers move along over the fixed rail. In the position shown in Fig. 2.6c, further motion of the slider causes the linkage to rotate counter-clockwise. The transfer rail falls downward and deposits the work carriers on the fixed rail in the next index position (Fig. 2.6d). The slider then moves to the left, which returns the transfer rail to its initial position. With this system, work carriers reaching the end of the assembly line are returned to the beginning of the line by a suitable conveyor. With the walking beam and all other transfer devices used on in-line machines, it is usual for each work carrier, after transfer, to be finally positioned and locked by a locating plunger before the assembly operation is initiated.

Another transfer system, known as *pawl transfer*, is shown in Fig. 2.7, where it can be seen that reciprocation of the transfer bar over a distance equal to the spacing of the workheads will cause the work carriers to index between the workheads.

The shunting work carrier transfer system is shown in Fig. 2.8. In this system, the work carriers have lengths equal to the distance moved during one index. Positions are available for work carriers at the beginning and end of the assembly line, where no assembly takes place. At the start of the cycle of operations, the work carrier position at the end of the line is vacant. A mechanism pushes the line of work carriers up to a stop at the end of the line and this indexes the work carriers one position. The piston then withdraws and the completed assembly at the end of the line is removed. The empty work carrier from a previous cycle that has been delivered by the return conveyor is raised into position at the beginning of the assembly line.

Although the system described here operates in the vertical plane, the return of work carriers can also be accomplished in the horizontal plane. In this case, transfer from the assembly line to the return conveyor (and vice versa) is simpler, but greater floor area is used. In practice, when operating in the horizontal plane, it is more usual to dispense with the rapid return conveyor and to fit further assembly heads and associated transfer equipment in its place (Fig. 2.9). This system has the disadvantage that access to the various workheads may be difficult.

A further disadvantage with all shunting work carrier systems is that the work carriers themselves must be accurately manufactured. For example, if an error of 0.025 mm were to occur on the length of

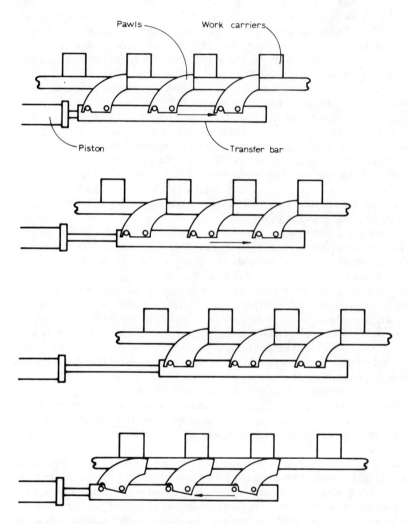

Fig. 2.7 Pawl-type transfer system.

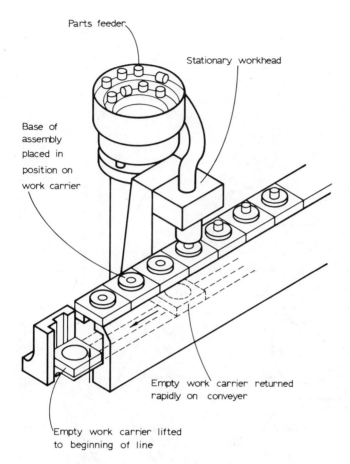

Parts feeder

Stationary workhead

Base of
assembly
placed in
position on
work carrier

Empty work carrier returned
rapidly on conveyer

Empty work carrier lifted
to beginning of line

Fig. 2. 8 In-line transfer machine with shunting work carriers returned
in vertical plane.

each work carrier in a 20-station machine, an error in alignment of
0.50 mm would occur at the last station. This error could create
serious difficulties in the operation of the workheads.

The chain-driven work carrier transfer system is shown in Fig. 2.10.
Basically, this machine uses an indexing mechanism that drives a chain
to which are attached the work carriers. The work carriers are spaced
to correspond to the distance between the workheads. (An alternative
to the chain drive is a flexible steel band.) With chain-driven indexing
systems, the problem of chain stretch must be considered in the design
of the machine. Clearly, if the chain stretches, the pitch of the work

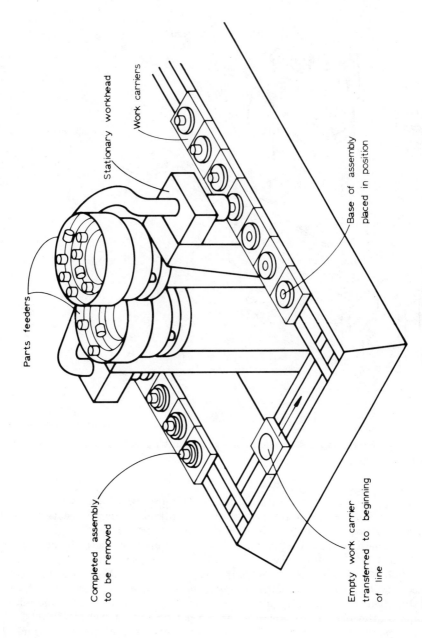

Parts feeders

Stationary workhead

Work carriers

Base of assembly
placed in position

Completed assembly
to be removed

Empty work carrier
transferred to beginning
of line

Fig. 2.9 In-line transfer machine with shunting work carriers returned in horizontal plane.

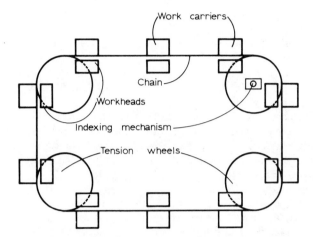

Fig. 2.10 Chain-driven transfer system.

carriers will vary. To overcome this fault, the usual method is to
arrange for the release of the chain tension on completion of the index
and to allow location plungers to position each work carrier relative
to the workhead. It must be ensured that the chain does not attempt
to index before the location plungers have been withdrawn.

Instead of attaching the work carriers rigidly to the chain, it is
possible to employ chain attachments which simply push the work
carriers along guides. In this case, the chain index can be arranged
to leave the work carriers short of their final position, allowing location
plungers to bring them into line with the workheads. With this system,
chain stretch does not present a serious problem.

2.3 Indexing Mechanisms

Huby* lists the factors affecting the choice of indexing mechanism for
an assembly machine as follows:

1. The required life of the machine
2. The dynamic torque capacity required
3. The static torque capacity required

---

*E. Huby, "Assembly Machine Transfer Systems," paper presented at
the Conference on Mechanized Assembly, July 1966, Royal College of
Advanced Technology, Salford, England.

4. The power source required to drive the mechanism
5. The acceleration pattern required
6. The accuracy of positioning required from the indexing unit

Generally, an increase in the size of a mechanism increases its life. Experience shows which mechanisms usually give longest life for given applications; this is discussed later.

The dynamic torque capacity is the torque that must be supplied by the indexing unit during the index of a fully loaded machine. The dynamic torque capacity is found by adding the effects of inertia and friction and multiplying by the life factor of the unit, the latter factor being found from experience with use of the indexing units.

The static torque capacity is the sum of the torques produced at the unit by the operation of the workheads. If individual location plungers are employed at each workhead, these plungers are usually designed to withstand the forces applied by the workheads; in this case, the static torque capacity required from the indexing unit will probably be negligible. The power required to drive an indexing unit will be obtained from the dynamic torque applied to the unit during the machine index.

The form of the acceleration curve for the indexing unit may be very important when there is any possibility that a partially completed assembly may be disturbed during the machine index. A smooth acceleration curve will also reduce the peak dynamic torque and will thus assist the driving motor in maintaining a reasonably constant speed during indexing and increasing the life of the machine. The accuracy of indexing required will not be great if locating plungers are employed to perform the final location of the work carriers or indexing table.

Various indexing mechanisms are available for use on automatic assembly machines, and typical examples are given in Figs. 2.11 through 2.13. These mechanisms fall into two principal categories: those that convert intermittent translational motion (usually provided by a piston) into angular motion by means of a rack and pinion or a ratchet and pawl (Fig. 2.11); those that are continuously driven, such as the Geneva mechanism (Fig. 2.12) or the crossover or scroll cam shown in Fig. 2.13.

For all but very low speed indexing or very small indexing tables, the rack and pinion or ratchet and pawl mechanisms are unsuitable, because they have a tendency to overshoot. The acceleration properties of both these systems are governed entirely by the acceleration pattern of the linear power source. To ensure a fairly constant indexing time, if the power source is a pneumatic cylinder, it is usual to underload the cylinder, in which case the accelerations at the beginning and end of the stroke are very high and produce undesirable shocks. The

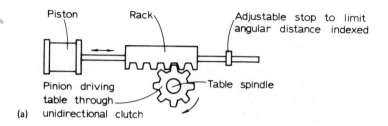

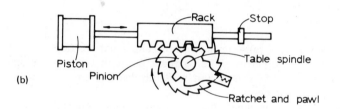

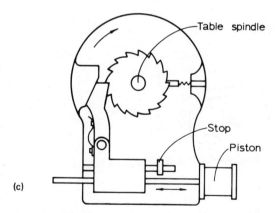

Fig. 2. 11 Indexing mechanisms: (a) rack and pinion with unidirectional clutch; (b) rack and pinion with ratchet and pawl; (c) ratchet and pawl.

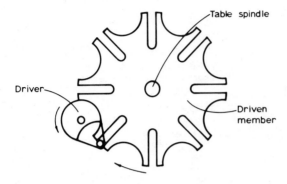

Fig. 2.12 Geneva mechanism.

ratchet and pawl requires a take-up movement and must be fairly
robust if it is to have a long life. The weakest point in the mechanism
is usually the pawl pin and if this is not well lubricated, the pawl will
stick and indexing will not occur.

The Geneva-type indexing mechanism has more general application
in assembly machines, but its cost is higher than the mechanisms described above. It is capable of transmitting a high torque relative to

Fig. 2.13 Crossover cam indexing unit.

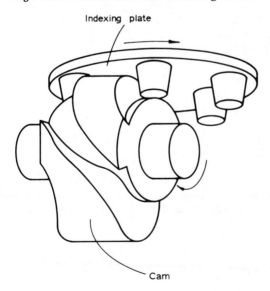

its size and has a smooth acceleration curve. However, it has a high peak dynamic torque immediately before and after the reversal from positive to negative acceleration. In its basic form, the Geneva mechanism has a fairly short life, but wear can be compensated for by adjustment of the centers. The weakest point in the mechanisms is the indexing pin, but breakages of this part can be avoided by careful design and the avoidance of undue shock reactions from the assembly machine. A characteristic of Geneva mechanisms is the jerk or rapid rate of change in acceleration at the beginning and end of each cycle. The main limitation in the use of the Geneva mechanism is its restriction in the number of stops per revolution. This is due primarily to the accelerations that occur with three-stop and more than eight-stop mechanisms.

In a Geneva mechanism, the smaller the number of stops, the greater is the adverse mechanical advantage between the driver and the driven members. This results in a high indexing velocity at the center of the indexing movement and gives a very peaked acceleration graph. On a three-stop Geneva this peaking becomes very pronounced, and since the mechanical advantage is very high at the center of the movement, the torque applied to the index plate is greatly reduced when it is most required. The solution to these problems results in very large mechanisms relative to the output torque available.

As the number of stops provided by a Geneva mechanism increases, the initial and final accelerations during indexing increase although the peak torque is reduced. This is due to the increased difficulty of placing the driver center close to the tangent of the indexing slot on the driven member.

For a unit running in an oil bath, the clearance between the driver and driven members during the locking movement is approximately 0.025 mm. To allow for wear in this region it is usual to provide a small center-distance adjustment between the two members. The clearance established after adjustments is the main factor governing the indexing accuracy of the unit, and this will generally become less accurate as the number of stops is increased. Because of the limitations in accuracy, it is usual to employ a Geneva mechanism in conjunction with a location plunger; in this case a relatively cheap and accurate method of indexing is obtained.

The crossover cam type of indexing mechanism shown in Fig. 2.13 is capable of transmitting a high torque, has a good acceleration characteristic, and is probably the most consistent and accurate form of indexing mechanism. Its cost is higher than the alternative mechanisms described above and it has the minor disadvantage that it is rather bulky. The acceleration characteristics are not fixed as with other types of indexing mechanism, but a crossover cam can be designed to give almost any required form of acceleration curve. The normal type of cam is designed to give a modified trapezoidal form of acceleration curve, which gives a low peak dynamic torque and a fairly low mean torque. The cam can be designed to give a wide range of stops per revolution of the index plate, and the indexing is inherently accurate. A further advantage is that it always has at least two indexing pins in contact with the cam.

 Figure 2.14 shows the acceleration patterns of the modified trapezoid, sine, and modified sine cams and the Geneva mechanism for the complete index of a four-stop unit. It can be seen that the modified trapezoidal form gives the best pattern for smoothest operation and lowest peaking. The sine and modified sine both give smooth acceleration, but the peak torque is increased, whereas with the Geneva mechanism, the slight initial shock loading and the peaking at the reversal of the acceleration are clearly evident.

Fig. 2.14 Comparison of acceleration curves for a Geneva mechanism and various designs of crossover cam: modified trapezoidal, ———; four-stop Geneva, — ——; modified sine, — - —; sine, ----. (Adapted from E. Huby, "Assembly Machine Transfer Systems," paper presented at the Conference on Mechanized Assembly, July 1966, Royal College of Advanced Technology, Salford, England.)

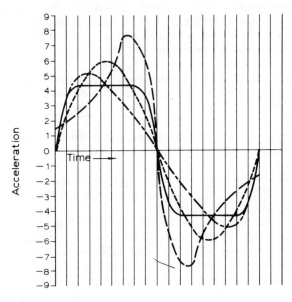

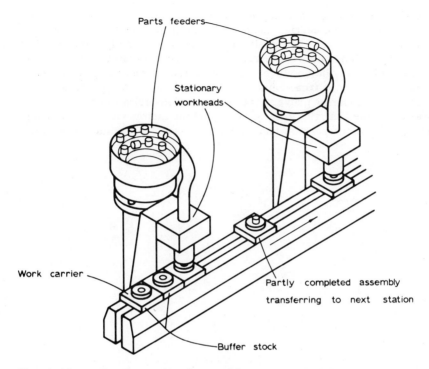

Parts feeders

Stationary
workheads

Work carrier

Partly completed assembly
transferring to next station

Buffer stock

Fig. 2.15 In-line free-transfer machine.

2.4 Operator-Paced Free-Transfer Machine

With all the transfer systems described earlier, it is usual for the
cycle of operations to occur at a fixed rate, and any manual operations
involved must keep pace. This is referred to as machine pacing.
Machines are available, however, where a new cycle of operations can
be initiated only when signals are received, indicating that all the
previous operations have been completed. This is referred to as
operator pacing.

   One basic characteristic that is common to all the systems described
is that a breakdown of any individual workhead will stop the whole
machine and production will cease until the fault has been cleared.
One type of in-line intermittent operator-paced machine, known as a
free-transfer machine (Fig. 2.15), does not have this limitation. In
this design, the spacing of the workstations is such that buffer stocks
of assemblies can accumulate between adjacent stations. Each workhead
or operator works independently and the assembly process is initiated

by the arrival of a work carrier at the station. The first operation is to lift the work carrier clear of the continuously moving chain conveyor and clamp it in position. After the assembly operation has been completed, the work carrier is released and transferred to the next station by the conveyor, provided that a vacant space is available. Thus, on a free-transfer machine, a fault at any one station will not necessarily prevent the other stations from working. It will be seen later that this can be an important factor when considering the economics of various transfer machines for automatic assembly.

Chapter 3

Vibratory Bowl Feeders

The vibratory bowl feeder is the most versatile of all hopper feeding devices for small engineering parts. In this feeder (Fig. 3.1) the track along which the parts travel is helical in form and passes around the inside wall of a shallow cylindrical hopper or bowl. The bowl is usually supported on sets of inclined leaf springs secured to a heavy base. Vibration is applied to the bowl from an electromagnet mounted on the base, and the support system constrains the movement of the bowl so that it has a torsional vibration about its vertical axis coupled with a linear vertical vibration. The motion is such that any part of the inclined track vibrates along a short-approximately straight path, which is inclined to the horizontal at an angle greater than that of the track. When component parts are placed in the bowl, the effect of the vibratory motion is to cause them to climb up the track to the outlet at the top of the bowl. Before considering the characteristics of vibratory bowl feeders, it is necessary to examine the mechanics of vibratory conveying. For this purpose it is convenient to deal with the motion of a part on a straight vibrating track that is inclined at a small angle to the horizontal.

3.1 Mechanics of Vibratory Conveying

In the following analysis, the track of a vibratory feeder is assumed to move bodily with simple harmonic motion along a straight path inclined at an angle $(\theta + \psi)$ to the horizontal as shown in Fig. 3.2. The angle of inclination of the track is θ and ψ is the angle between the track and its line of vibration. The frequency of vibration f (usually 60 Hz in practice) is conveniently expressed in this analysis as $\omega = 2\pi f$

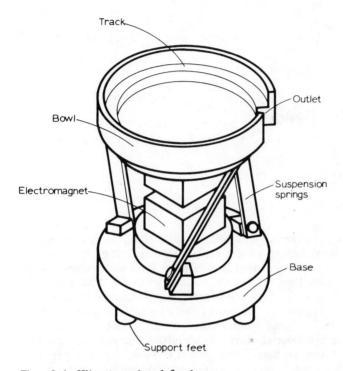

Fig. 3.1 Vibratory bowl feeder.

Fig. 3.2 Force acting on a part in vibratory feeding.

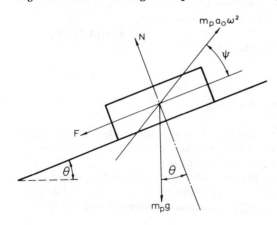

rads, where ω is the angular frequency of vibration. The amplitude of vibration a_0 and the instantaneous velocity and acceleration of the track may all be resolved in directions parallel and normal to the track. These components will be referred to as parallel and normal motions and the normal motions will be indicated by the suffix n.

It is assumed in the analysis that the motion of a part of mass m_p is independent of its shape and that air resistance is negligible. It is also assumed that there is no tendency for the part to roll down the track.

It is useful to consider the behavior of a part that is placed on a track whose amplitude of vibration is increased gradually from zero. For small aplitudes the part will remain stationary on the track because the parallel inertia force acting will be too small to overcome the frictional resistance F between the part and the track. Figure 3.2 shows the maximum inertia force acting on the part when the track is at the upper limit of its motion. This force has parallel and normal components of $m_p a_0 \omega^2 \cos \psi$ and $m_p a_0 \omega^2 \sin \psi$, respectively, and it can be seen that for sliding up the track to occur,

$$m_p a_0 \omega^2 \cos \psi > m_p g \sin \theta + F \qquad (3.1)$$

where

$$F = \mu_s N = \mu_s (m_p g \cos \theta - m_p a_0 \omega^2 \sin \psi) \qquad (3.2)$$

and where μ_s is the coefficient of static friction between the part and the track. The condition for forward sliding up the track to occur is, therefore, given by combining Eqs. (3.1) and (3.2). Thus,

$$\frac{a_0 \omega^2}{g} > \frac{\mu_s \cos \theta + \sin \theta}{\cos \psi + \mu_s \sin \psi} \qquad (3.3)$$

Similarly, it can be shown that for backward sliding to occur during the vibration cycle,

$$\frac{a_0 \omega^2}{g} > \frac{\mu_s \cos \theta - \sin \theta}{\cos \psi - \mu_s \sin \psi} \qquad (3.4)$$

The operating conditions of a vibratory conveyor may be expressed in terms of the dimensionless normal track acceleration A_n/g_n, where A_n is the normal track acceleration ($A_n = a_n \omega^2 = a_0 \omega^2 \sin \psi$), g_n the normal acceleration due to gravity (= $g \cos \theta$), and g the acceleration due to gravity (= 9.81 m/s^2). Thus,

$$\frac{A_n}{g_n} = \frac{a_0 \omega^2 \sin \psi}{g \cos \theta} \qquad (3.5)$$

Substitution of Eq. (3.5) in Eqs. (3.3) and (3.4) gives:

for forward sliding:

$$\frac{A_n}{g_n} > \frac{\mu_s + \tan \theta}{\cot \psi + \mu_s} \qquad (3.6)$$

for backward sliding:

$$\frac{A_n}{g_n} > \frac{\mu_s - \tan \theta}{\cot \psi - \mu_s} \qquad (3.7)$$

For values of $\mu_s = 0.8$, $\theta = 3$ degrees (0.05 rad) and $\psi = 30$ degrees (0.52 rad), Eqs. (3.6) and (3.7) show that the ratio A_n/g_n must be greater than 0.34 for forward sliding to occur and greater than 0.8 for backward sliding. With these conditions it is clear that, for all amplitudes of vibration giving a value of A_n/g_n greater than 0.34, forward sliding will predominate and the part will climb the track, sliding forward or both forward and backward during each vibration cycle.

The limiting condition for forward conveying to occur is given by comparing Eqs. (3.6) and (3.7). Thus, for forward conveying

$$\tan \psi > \frac{\tan \theta}{\mu_s^2}$$

or, when θ is small,

$$\tan \psi > \frac{\theta}{\mu_s^2} \qquad (3.8)$$

For values of $\mu_s = 0.8$ and $\theta = 3$ degrees (0.05 rad), ψ must be greater than 4.7 degrees (0.08 rad) for forward conveying to occur.

For sufficiently large vibration amplitudes the part will leave the track and "hop" forward during each cycle. The condition for this to occur is where the normal reaction N between the part and the track becomes zero.

From Fig. 3.2,

$$N = m_p g \cos \theta - m_p a_0 \omega^2 \sin \psi \qquad (3.9)$$

and, therefore, for the part to leave the track

$$\frac{a_0 \omega^2}{g} > \frac{\cos \theta}{\sin \psi}$$

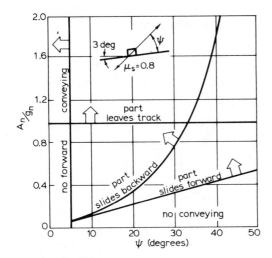

Fig. 3.3 Limiting conditions for various modes of vibratory conveying. A_n is the normal track acceleration, g_n the normal gravitational acceleration.

or

$$\frac{A_n}{g_n} > 1.0 \qquad\qquad (3.10)$$

It is clear from the earlier examples, however, that the part would slide forward before it leaves the track during each cycle. Figure 3.3 graphically illustrates these equations. This shows the effect of the vibration angle ψ on the limiting values of the dimensionless normal acceleration A_n/g_n for forward sliding to occur, for both forward and backward sliding to occur, and for the part to hop along the track.

The detailed types of motion that may occur in vibratory feeding have been described in the literature.* For all conditions, the part starts to slide forward at some instant when the track is nearing the upper limit of its motion. When there is no hopping mode, this forward sliding continues until the track is nearing the lower limit of its motion, at which point the part may remain stationary, relative to the track, or slide backward until the cycle is complete. In some cases, the stationary period is followed by a period of backward sliding only or backward

*A. H. Redford and G. Boothroyd, "Vibratory Feeding," *Proc. I Mech. Eng.*, vol. 182, part 1, no. 6, 1967-1968, p. 135.

sliding followed by yet another stationary period. Finally, the forward sliding is followed by a period of backward sliding and then a stationary period to complete the cycle.

Analysis and experiment have shown that higher feed rates are obtained with the hopping mode of conveying (that is, when $A_n/g_n >$ 1.0). The modes of conveying are summarized in the following flow diagram:

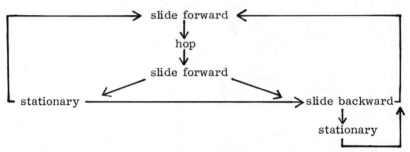

Clearly, a complete analysis of all the possible modes of vibratory conveying is complicated. Such an analysis has been made* and leads to equations that must be solved numerically with the aid of a digital computer. For the purposes of the present discussion it is considered adequate to describe only the main results of this analysis and the results of some experimental tests. In the following, the effects of frequency f, track acceleration A_n/g_n, track angle θ, vibration angle ψ, and the effective coefficient of friction μ on the mean conveying velocity v_m are discussed separately.

3.2 Effect of Frequency

One principal result of the theoretical work is that for given conditions and for constant track acceleration (that is, A_n/g_n is constant), the mean conveying velocity v_m is inversely proportional to the vibration frequency f. Hence,

$$fv_m = \text{constant} \qquad (3.11)$$

This is illustrated in Fig. 3.4, where the effect of track acceleration on the mean conveying velocity is plotted for three values of the vibration angle ψ. It can be seen that the experimental points for a range of frequencies fall on one line when the factor fv_m is used as a measure of the conveying velocity. This verifies the prediction of the

---

*Ibid.

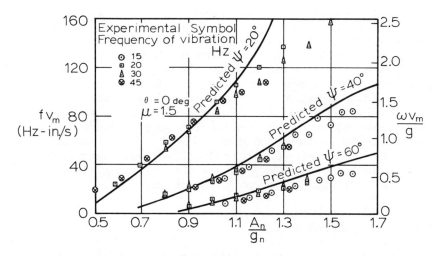

Fig. 3.4 Effect of vibration angle, track acceleration, and frequency on conveying velocity, where ψ is the vibration angle (degrees), f the frequency (Hz), θ the track angle (degrees), μ the coefficient of friction, and v_m the mean conveying velocity. (From A. H. Redford and G. Boothroyd, "Vibratory Feeding," *Proc. I Mech. Eng.*, vol. 182, part 1, no. 6, 1967-1968, p. 135.)

theoretical analysis. One consequence of this result is that for high conveying velocities and hence high feed rates, it is desirable to use as low a frequency as practicable. However, since the track accelerations must be kept constant, this result means a corresponding increase in track amplitude. The mechanical problems of connecting the feeder to a stationary machine imposes a lower limit on the frequency. The results of experiments* in England have shown that some advantage is to be gained from lowering the operating frequency of a bowl feeder from the usual 50 Hz to 25 Hz.

3.3 Effect of Track Acceleration

Figure 3.4 shows that an increase in track acceleration A_n/g_n generally produces an increase in conveying velocity. At some point, however, although the theoretical analysis predicts further increases in velocity, increases in A_n/g_n cease to have a significant effect. This finding may be explained as follows.

---

*Ibid.

If the track acceleration is increased until $A_n/g_n > 1.0$, the part starts to hop once during each cycle as described earlier. At first, the velocity of impact as the part lands on the track is small but, as the track acceleration is increased further, the impact velocity also increases until, at some critical value, the part starts to bounce. Under these circumstances the feeding cycle becomes erratic and unstable and the theoretical predictions are no longer valid.

To obtain the most efficient feeding conditions, it is necessary to operate with values of A_n/g_n greater than unity but below the values that will produce unstable conditions. From Fig. 3.4 it can be seen that within this range, an approximately linear relationship exists between the factors fv_m and A_n/g_n for each value of ψ and for given values of track angle, θ, and coefficient of friction, μ.

3.4 Effect of Vibration Angle

From Fig. 3.4 it can be seen that the conveying velocity is sensitive to changes in the vibration angle ψ. The effect is shown more clearly in Fig. 3.5, which indicates that an optimum vibration angle exists for given conditions. For clarity, these theoretical predictions are shown without supporting experimental evidence. Previous work[*] has resulted in the relationship between optimum vibration angle ψ_{opt} and

---

[*]Ibid.

Fig. 3.5 Theoretical results showing the effect of vibration angle on the mean conveying velocity. (From A. H. Redford and G. Boothroyd, "Vibratory Feeding," *Proc. I Mech. Eng.*, vol. 182, part 1, no. 6, 1967-1968, p. 135.)

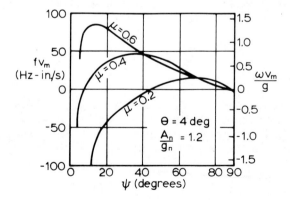

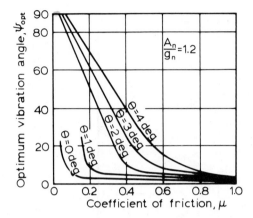

Fig. 3.6 Theoretical results showing the effect of coefficient of friction on the optimum vibration angle. (From A. H. Redford and G. Boothroyd, "Vibratory Feeding," *Proc. I Mech. Eng.*, vol. 182, part 1, no. 6, 1967-1968, p. 135.)

coefficient of friction as shown in Fig. 3.6 for a practical value of track acceleration where A_n/g_n is 1.2.

3.5 Effect of Track Angle

Figure 3.7 shows the effect of track angle θ on the conveying velocity for various track accelerations when μ is 0.2. These results show that the highest velocities are always achieved when the track angle is zero and second, that forward conveying is obtained only with small track angles. The mechanical design of a bowl feeder necessitates a positive track angle of three or four degrees in order to raise the parts to the bowl outlet. However, it can be seen from the figure that even if conveying can be achieved on the track, the mean conveying velocity will be significantly lower than that around the flat bottom of the bowl. This means that in practice, the parts on the track will invariably be pushed along by those in the bottom of the bowl, which will tend to circulate at a greater speed. This leads to certain problems in the design of the orienting devices which are generally placed around the upper part of the bowl track. During the testing of such orienting devices, parts transported individually along the track may behave correctly. However, when the bowl is filled with parts and a line forms along the track, the parts tend to be forced through the orienting devices by the pressure of those in the bottom of the bowl. This

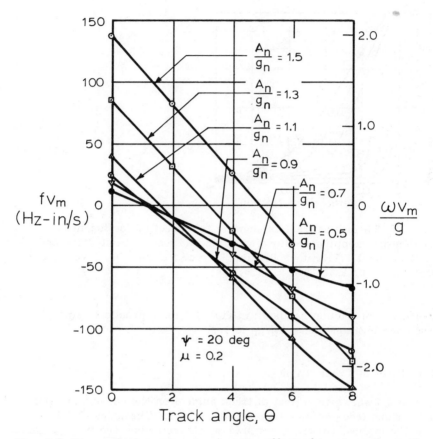

Fig. 3.7 Theoretical results showing the effect of track angle on the conveying velocity. (From A. H. Redford, "Vibratory Conveyors," Ph.D. thesis, Salford University, Salford, England, 1966.)

pressure may often lead to jamming and general unreliability in operation.

From the foregoing discussion it is clear that when considering the unrestricted feed rate from a bowl feeder, a track angle of zero degrees should be employed because the feeding characteristics in the flat bowl bottom will generally govern the overall performance of the feeder.

3.6 Effect of Coefficient of Friction

The practical range of the coefficient of friction in vibratory feeding is from 0.2 to 1.0. The figure of 0.2 is representative of a steel part conveyed on a steel track. By lining the track with rubber, a common practice in industry, the coefficient of friction may be raised to approximately 0.8.

Figure 3.8 shows the effect of the coefficient of friction on the conveying velocity for a horizontal track, a vibration angle of 20 degrees, and for various track accelerations. It can be seen that for practical values of track acceleration, an increase in friction leads to an increase in conveying velocity, hence the advantage of increasing friction by coating the tracks of bowl feeders with rubber. Coatings can also reduce the noise level due to the motion of the parts, a consideration that is often of paramount importance.

3.7 Estimating the Mean Conveying Velocity

At any point on a horizontal track, the ratio of the amplitudes of the vertical and horizontal components of vibration is equal to the tangent of the vibration angle ψ. When the bowl is operating properly with no

Fig. 3.8 Theoretical results showing the effect of the coefficient of friction on the conveying velocity. (From A. H. Redford, "Vibratory Conveyors," Ph.D. thesis, Salford University, Salford, England, 1966.)

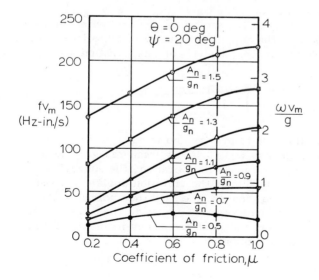

rocking motion, the vertical component of motion a_n will be the same at every location in the bowl. The magnitude of the horizontal component a_p, however, changes with radial position.

The horizontal component increases linearly with increasing radial position. If the vibration angle ψ_1 at radial position r_1 is known, the vibration angle ψ_2 at a radial position r_2 is is found from

$$\tan \psi_2 = \frac{r_1}{r_2} \tan \psi_1 \qquad (3.12)$$

If the leaf springs are inclined at 60 degrees (1.05 rad) from the horizontal plane and attached to the bowl 100 mm from the bowl center, the vibration angle at this radius is the complement of the spring inclination angle or 30 degrees (0.52 rad) if the vibration of the base is neglected. If the vibration of the base is important, the vibration angle should be determined experimentally by comparing the signals from two accelerometers: one mounted vertically, the other horizontally. The value of the vibration angle at a radial position 150 mm from the bowl center can be found from Eq. (3.12) and is

$$\psi_2 = \arctan \left(\frac{100}{150} \tan 30°\right) = 21° \ (0.37 \text{ rad}) \qquad (3.13)$$

The vibratory motion of a bowl feeder causes parts, randomly deposited in the bottom of the bowl, to climb the helical track on the interior of the bowl wall. The conveying velocity of the parts on the inclined track is usually governed by the pushing action of the parts circulating around the bottom of the bowl. For those parts moving on the horizontal bottom of the bowl, the conveying velocity v_m depends mainly on the vibration angle ψ, the amplitude of vibration a_0, and the frequency of vibration f or the angular frequency of vibration ω, where ω equals $2\pi f$. A simple dimensional analysis of this situation shows that

$$\frac{v_m \omega}{g} = \text{function} \left[\left(\frac{a_0 \omega^2 \sin \psi}{g} \right), \ (\psi) \right] \qquad (3.14)$$

where g is the acceleration due to gravity and is equal to 9.81 m/s^2. The functional relationship from Eq. (3.14) is presented graphically in Fig. 3.9. For the usual case of 60-Hz vibration, the conveying velocity is also shown as a function of the vibration angle ψ and the vertical amplitude of vibration, that is, the amplitude normal to the horizontal track. The dimensionless scales shown to the top and right in Fig. 3.9 can be used for any vibration frequency including 60 Hz, but this requires additional computation.

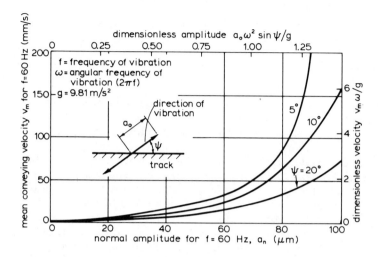

Fig. 3.9 Estimation of the mean conveying velocity on a horizontal track.

For example, suppose that a 60-Hz vibratory bowl feeder has the start of the track, on the bottom of the bowl, located 150 mm from the bowl center, and the vibration angle at this point is 21 degrees (0.37 rad). Then according to Fig. 3.9, the conveying velocity v_m is approximately 35 mm/s when a_n is 80 μm. As a comparison, if a_n or $a_0 \sin \psi$ equals 80 μm, the value of the dimensionless amplitude

$$\frac{a_0 \omega^2 \sin \psi}{g} = \frac{(80 \times 10^{-6})(2\pi 60)^2}{9.81} = 1.16 \qquad (3.15)$$

From Fig. 3.9, if ψ equals 21 degrees (0.37 rad), the dimensionless velocity $v_m \omega / g$ equals 1.35 and v_m equals 35 mm/s. The corresponding value of the horizontal amplitude a_p is 210 μm.

The vibration amplitude is usually set while observing a special decal mounted on the outer rim of the bowl. This decal is used to measure the peak-to-peak amplitude or twice the horizontal amplitude of vibration at that point. The correct value of this horizontal amplitude depends on the bowl diameter and is found from geometry.

Using the previous example on a bowl 600 mm in diameter, the parallel amplitude at the rim is

$$\frac{300}{150} (210) = 420 \text{ μm} \qquad (3.16)$$

so that the peak-to-peak setting is 0.84 mm. As a consequence, the
vibration angle for the last horizontal section of the track is

$$\arctan\left(\frac{80}{420}\right) = 11° \ (0.19 \ \text{rad}) \tag{3.17}$$

As the parts leave the bowl their conveying velocity is now 70 mm/s,
from Fig. 3.9.

Although parts are apparently conveyed by vibratory motion with
an almost constant conveying velocity, this motion is, in actuality, a
combination of a variety of dissimilar smaller motions giving the total
effect of smooth translation. This combination of smaller motions is
cyclic and usually repeats with the frequency of the drive. Some of
the details of this motion are important in the design of orienting
devices used in vibratory bowl feeders. Figures 3.10 and 3.11 show
the effective length J and height H of a hop for a point mass traveling
on a horizontal track as shown in Fig. 3.12. The effective length of
the hop is the smallest gap in the track that will reject all point masses
traveling with this motion.

The magnitude of this effective hop can be determined from Fig. 3.10.
If the normal amplitude a_n is 80 μm for 60-Hz vibration and the vibration angle ψ is 11 degrees (0.19 rad), then from this figure the value
of the effective hop J is 0.6 mm. Scales on the top and right side of
Fig. 3.10 can be used for other frequencies of vibration, as explained
in the discussion of Fig. 3.9.

Fig. 3.10 Theoretical estimate of the length of the effective hop on a
horizontal track.

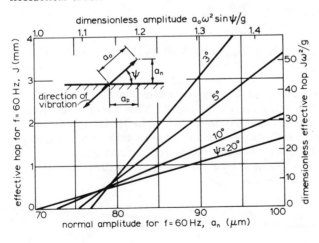

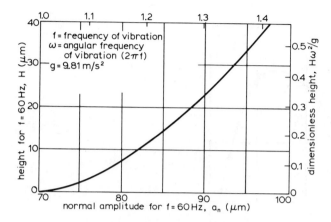

Fig. 3.11 Theoretical estimate of the maximum height reached by a hopping part above a horizontal track.

Similarly, Fig. 3.11 can be used to determine the magnitude of the maximum height H of the hop above the horizontal track. Using the previous conditions (a_n = 80 μm and 60 Hz), H is 8 μm.

3.8 Load Sensitivity

One of the main disadvantages of vibratory bowl feeders is their change in performance as the bowl gradually empties. This change occurs because, for a constant power input, the amplitude of vibration, and hence the maximum bowl acceleration, usually increases as the effective mass of the loaded bowl reduces. It can be deduced from Fig. 3.13 that this increase in bowl acceleration will generally result in an increase in the unrestricted feed rate. Vibratory bowl feeders are often

Fig. 3.12 Typical part motion, including the hop.

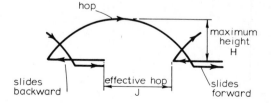

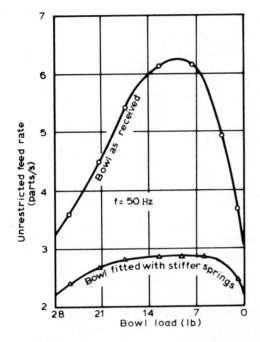

Fig. 3.13 Experimentally determined load sensitivity of a commercial bowl feeder. (From A. H. Redford, "Vibratory Conveyors," Ph.D. thesis, Salford University, Salford, England, 1966.)

Fig. 3.14 Effect of bowl load on bowl acceleration. (From A. H. Redford, "Vibratory Conveyors," Ph.D. thesis, Salford University, Salford, England, 1966.)

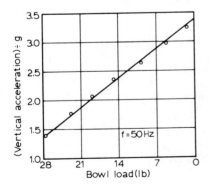

used to convey and orient parts for automatic assembly and, since the workheads on an assembly machine are designed to work at a fixed cycle time, the parts can only leave the feeder at a uniform rate. The feeder must therefore be adjusted to overfeed slightly under all conditions of loading, and excess parts are continuously returned from the track to the bottom of the bowl.

The change in performance as a feeder gradually empties is referred to as its load sensitivity and the upper curve in Fig. 3.13 shows how the unrestricted feed rate for a commercial bowl feeder in the as-received condition varied as the bowl emptied. It can be seen that the maximum feed rate occurred when the bowl was approximately 25% full and that this represented an increase of approximately 100% on the feed rate obtained with the bowl full. It is of interest to compare this result with the measured changes in bowl acceleration shown in Fig. 3.14, where it can be seen that the bowl acceleration, and hence the amplitude, increased continuously until the bowl became empty. Clearly, when a feeder empties, the feed rate will reduce to zero, but Fig. 3.13 shows that the feed rate begins to reduce much sooner than might be expected from Fig. 3.14. This behavior is considered to be due to the greater velocity of parts in the flat bowl bottom than that on the track; this was described earlier. When the bowl is full, the feed rate depends mainly on the feeding characteristics in the bottom of the bowl, where the general circulation of parts pushes those on the track. However, when the bowl empties sufficiently so that its contents are mainly held on the track, the pushing action ceases and the feed rate depends on the conveying velocity on the inclined track, which is generally lower than that on a horizontal surface. This explains the difference in character between the graphs in Figs. 3.13 and 3.14.

Figure 3.13 suggests that under the test conditions the as-received bowl feeder could be used to feed a workhead operating at a maximum rate of 3 cycles per second and that in this case there would be considerable recirculation of parts due to overfeeding. Assuming that the feeder is to be refilled when it becomes 25% full, the feeding characteristics between refills may be reasonably represented by a feed rate increasing linearly as the bowl empties. An analysis is now presented that may be used to estimate the unnecessary recirculation of parts in a bowl feeder with this characteristic.

3.9 Effect of Load Sensitivity on Recirculation of Parts

For the purpose of the following analysis, the load sensitivity S of a bowl feeder will be defined as the proportional change in unrestricted feed rate between refills. Figure 3.15 shows the relationship of parame-

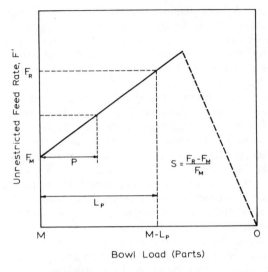

Fig. 3.15 Simple model of vibratory bowl feeder load sensitivity.

ters used in this analysis and the model is based on the results shown in Fig. 3.14. The maximum bowl load or bowl capacity is assumed to be M parts, and after L_p parts have been fed from the bowl, a refill of L_p parts will be added to the bowl, bringing the bowl load back again to M parts. It is also assumed that between refills the unrestricted feed rate F' increases linearly with reductions in bowl loading as shown in Fig. 3.15. Thus,

$$F' = F_m + SF_m \frac{P}{L_p} \qquad (3.18)$$

where F_m is the feed rate for a full bowl and P the number of parts fed from the bowl in time t after a refill. On an assembly machine, the bowl feeder delivery rate is constant and for convenience is assumed to be equal to F_m, the minimum unrestricted feed rate. Thus, the recirculation rate, which is the rate at which parts return to the bottom of the bowl, is

$$SF_m \frac{P}{L_p} = \frac{SF_m^2 t}{L_p} \qquad (3.19)$$

since P is equivalent to $F_m t$. The number of parts returned to the bottom of the bowl in a small time interval Δt is

$$\frac{SF_m^2 t}{L_p} \Delta t$$

However, it is not the number of parts recirculated that is important but the cumulative effect of recirculation of a single part, which, in general terms, is the number of parts recirculated divided by the number of parts in the bowl. Since for that same time interval Δt there are $(M - F_m t)$ parts in the bowl, the instantaneous average recirculation per part $\Delta R'$ equals the number of parts returned divided by the total number available, or

$$\Delta R' = \frac{SF_m^2 t}{L_p(M - F_m t)} \Delta t \qquad (3.20)$$

The effective average recirculation of a single part R' is the integral of Eq. (3.20) over the time interval L_p/F_m, which is the time between refills. Thus,

$$R' = \int_0^{L_p/F_m} \frac{SF_m^2 t}{L_p(M - F_m t)} dt$$

$$= S\left[-1 - \frac{M}{L_p} \ln\left(1 - \frac{L_p}{M}\right)\right] \qquad (3.21)$$

For example, if the feeder used in Fig. 3.14 is refilled when the bowl is half empty, which occurs near where the curve peaks, then

$$S \simeq \frac{6 - 3}{3} = 1$$

$$\frac{L_p}{M} = \frac{28 - 14}{28} = 0.5$$

and

$$R' = 1[-1 - 2 \ln (0.5)] = 0.39$$

This means that basically one of every three parts has been returned to the bottom of the bowl because of overfeeding. This does not include the effect of the orienting devices, which also reject improperly oriented parts back to the bottom of the bowl.

If the same feeder is refilled twice as often, then

$$S \simeq \frac{4.5 - 3}{3} = 0.5$$

$$\frac{L_p}{M} = \frac{28 - 21}{28} = 0.25$$

and

$$R' = 0.5[-1 - 4 \ln (0.75)] = 0.08$$

which shows a significant reduction in recirculation. This also shows that recirculation is sensitive to the refilling rate, with recirculation increasing as the bowl empties.

3.10 Solutions to Load Sensitivity

One of the simplest solutions to load sensitivity and the one most commonly used is the load detector switch together with a secondary feeder. The load detector switch is simply a mechanical arm and limit switch which detects when the level of parts in the bottom of the bowl falls below some predetermined level. When closed, the switch activates the secondary feeder and refills the bowl to the predetermined level. This action essentially increases the frequency of refills, reducing the recirculation effect to near zero.

A second solution requires modification to the feeder. Frequency response curves for the vibratory bowl feeder used in the previous experiments are presented in Fig. 3.16. These curves show the effect of changes in the forcing frequency on the bowl acceleration for a constant power input and for various bowl loadings. In these tests the power input is less than that employed for the results in Fig. 3.14, but they show the same effect. For a forcing frequency of 50 Hz, the maximum bowl acceleration is sensitive to changes in bowl loading. However, it can also be seen that for a forcing frequency of approximately 44 Hz, the bowl acceleration is approximately constant for all bowl loadings. Under these conditions the load sensitivity would be considerably reduced. Alternatively, if the spring stiffness of the bowl supports were to be increased sufficiently, it is clear that this would have the effect of shifting the response curves so that the changes in bowl acceleration would be minimized for a forcing frequency of 50 Hz. The natural frequency of the empty as-received bowl was approximately 53 Hz, and tests showed that if this was increased to 61 Hz by increasing the support spring stiffness, the load sensitivity of the feeder was considerably reduced. The lower curve in Fig. 3.13 shows this effect. It can also be seen, however, that the feed rates

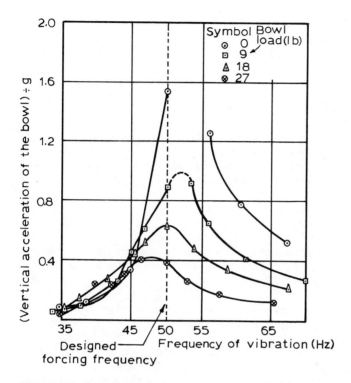

Fig. 3.16 Frequency response curves for a vibratory bowl feeder, showing the effect of bowl load. (From A. H. Redford, "Vibratory Conveyors," Ph.D. thesis, Salford University, Salford, England, 1966.)

have been reduced by stiffening the support springs and therefore, in order to maintain the higher feed rate, a more powerful drive would be required. Generally, vibratory bowl feeders are tuned to have a natural frequency just slightly higher than the frequency of the drive, to minimize the power by utilizing the ease of transmitting vibration at or near the natural frequency.

A third solution uses on/off controls in the feed track or delivery chute to control the operation of the feeder. A line of parts is stored in the external feed track, and when the line becomes small the lower sensor activates the feeder, filling the line to the upper sensor, which in turn shuts the feeder off. This has the effect of reducing the value of S to zero and eliminating recirculation.

Some of the more expensive vibratory bowl feeders use silicon-controlled rectifier (SCR) drive systems which can be coupled with accelerometer feedback to hold the vibration amplitude constant. This keeps the mean conveying velocity of the parts from increasing as the bowl empties and essentially makes S, the overfeeding parameter, equal to zero.

3.11 Orientation of Parts

One of the principal reasons for the wide application of the vibratory bowl feeder is its ability to feed and orient a large majority of the small parts used in engineering assembly work. The various methods of orienting parts are described in Chapter 5.

3.12 Bowl Feeder Design

The following gives a summary of the results and conclusions obtained in the work described in this chapter, which would be useful in the design of a vibratory bowl feeder:

1. For a given track acceleration, the mean conveying velocity in a vibratory bowl feeder is inversely proportional to the operating frequency.
2. For a given frequency, an increase in vibration amplitude (or track acceleration) will increase the conveying velocity. However, with large accelerations the behavior of the parts becomes erratic and unpredictable.
3. For maximum conveying velocity, the track angle should be as small as possible. However, feeding around the flat bowl bottom will always be faster than on the inclined track, and therefore the parts will generally be pushed up the track by the circulation of those on the bottom.
4. An optimum vibration angle exists for any given conditions.
5. Higher coefficients of friction between the parts and the track will generally give higher conveying velocities.
6. For any bowl feeder, a support spring stiffness may be chosen that will give a reasonably low load sensitivity. However, with many commercial feeders, this will necessitate an increase in the power from the electromagnet in order to maintain a sufficiently high feed rate.

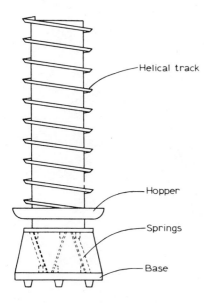

Fig. 3.17 Spiral elevator.

3.13 Spiral Elevators

A device commonly employed for elevating and feeding component parts is the spiral elevator. A typical spiral elevator is illustrated in Fig. 3.17; it can be seen that the drive is identical to that used for a vibratory bowl feeder. The helical track passes around the outside of a cylindrical tube. This device is not generally used to orient parts because the parts cannot readily be rejected back into the hopper bowl situated at the base of the elevator. Since the mode of conveyance for the parts is identical to that obtained with a vibratory bowl feeder, the results and discussion presented above and the design recommendations made will also apply to the spiral elevator.

Chapter 4

Nonvibratory Feeders

Although the vibratory bowl feeder is the most widely employed and most versatile parts feeding device, many other types of parts feeders are available. Usually these are only suitable for feeding certain basic types of component parts, but when feeding these parts, better results may be obtained for a smaller capital outlay with feeders other than the vibratory type.

One point that must be borne in mind when considering parts feeders is that, in automatic assembly, the output of parts from the feeder is always restricted by the machine being fed. The machine will generally use parts at a strictly uniform rate and this may be referred to as the machine rate. In the design and testing of parts feeders, it is often convenient to observe the feed rate when the feeder is not connected to a machine, that is, when no restriction is applied to the output of the feeder. The feed rate under these circumstances will be referred to as the unrestricted feed rate. Clearly, in practice the mean unrestricted feed rate must not fall below the machine rate.

Certain other general requirements of parts feeders may be summarized as follows: The unrestricted feed rate should not vary widely because this will simply mean that when the feeder is connected to a machine, the parts will be continuously recirculated within the feeder for much of the time. This will cause excessive wear and may eventually damage the parts. This undesirable characteristic often occurs in parts feeders where the feed rate is sensitive to changes in the quantity of parts present in the feeder and will be referred to as the load sensitivity of the feeder.

With parts feeders suitable for automatic machines it is necessary that all the parts be presented to the machine in the same orientation; that is, they must be fed correctly oriented. Some feeders are able

to feed and orient many types of part, whereas others are only able
to handle a very limited range of part shapes.

Undoubtedly, the reliability of a parts feeder is one of its most
important characteristics. Parts feeders should be designed so that
the possibility of parts jamming in the feeder, or in its orienting
devices, is minimized or eliminated.

It is sometimes suggested that parts feeders can also act as inspection devices. It is possible to design certain parts feeders so that
misshapen parts, swarf, and so on, will not be fed to the machine but
will be rejected by the device fitted to the feeder. This can be an
important feature because defective parts or foreign matter, if fed to
the machine, will probably cause a breakdown and may stop the whole
production line.

Some parts feeders are noisy in operation and some tend to damage
certain types of part. Obviously, both these aspects of parts feeding
must be considered when studying the possible alternatives for a
particular application.

Parts feeders can generally be classified into the following: reciprocating feeders; rotary feeders; belt feeders; and vibratory feeders,
which were discussed in Chapter 3. A selection of the more common
feeding devices within each of these groups will now be described and
discussed.

4.1 Reciprocating Tube Hopper Feeder

A reciprocating tube hopper is illustrated in Fig. 4.1 and consists
of a conical hopper with a hole in the center through which a delivery
tube passes. Relative vertical motion between the hopper and the

Fig. 4.1 Reciprocating tube hopper.

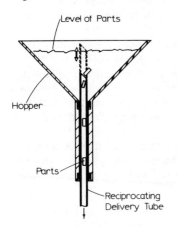

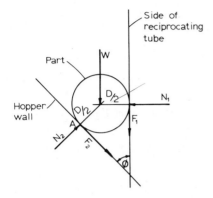

Fig. 4.2 Forces acting on part jammed between hopper wall and tube.

tube is achieved by reciprocating either the tube or the hopper.
During the period when the top of the tube is below the level of parts,
some parts will fall into the delivery tube. It is usual to machine the
top of the tube at an angle so that a part resting across the opening
will fall clear and not block the opening as the tube is pushed upward
through the mass of parts. Care must be taken in choosing the angle
of the conical hopper because if the angle is too small, there is a possibility of parts jamming between the tube and the hopper.

Figure 4.2 shows the forces acting on a cylindrical part jammed in
this way when the tube is moving downward relative to the hopper.
The force W acting vertically downward represents the weight of the
part together with any additional force that may be present due to
parts resting on top of the one shown. Resolving forces vertically
and horizontally and taking moments about A gives

$$F_1 + W + F_2 \cos \phi = N_2 \sin \phi \qquad (4.1)$$

$$N_1 = N_2 \cos \phi + F_2 \sin \phi \qquad (4.2)$$

$$F_1(1 + \cos \phi) \frac{D}{2} + W \left(\frac{D}{2}\right) \cos \phi = N_1 \left(\frac{D}{2}\right) \sin \phi \qquad (4.3)$$

where ϕ is the hopper wall angle and D is the diameter of the part.
Eliminating W from Eqs. (4.1) and (4.3) gives, after rearrangement,

$$N_1 \sin \phi = F_1(1 + \cos \phi) + \cos \phi (N_2 \sin \phi - F_2 \cos \phi - F_1) \qquad (4.4)$$

The maximum value of F_2 is given by $\mu_s N_2$ (where μ_s is the coefficient of static friction), and thus writing $F_2 = \mu_s N_2$ in Eqs. (4.2) and (4.4) and eliminating N_2 gives

$$\frac{F_1}{N_1} = \frac{\mu_s}{\cos \phi + \mu_s \sin \phi} \qquad (4.5)$$

For the tube to slide, $F_1/N_1 > \mu_s$, and therefore from Eq. (4.5),

$$\frac{1}{\cos \phi + \mu_s \sin \phi} > 1 \qquad (4.6)$$

Expression (4.6) indicates that the value of ϕ should be as large as possible to prevent jamming when μ_s is large. However, when $\mu_s <$ cot ϕ, the parts cannot slide down the hopper wall. The best compromise is probably given by writing the limiting conditions:

$$\mu_s = \cot \phi \qquad (4.7)$$

$$\cos \phi + \mu_s \sin \phi = 1 \qquad (4.8)$$

Combining Eqs. (4.7) and (4.8) gives $\phi = 60$ degrees, and on substituting this value in expression (4.6) it is found that, to prevent jamming under these conditions, the coefficient of friction μ_s must be less than 0.577. Since this value is greater than that expected in practice, it may be concluded that with a hopper angle of 45 degrees, the possibility of jamming will generally be avoided if the coefficient of friction is less than 0.414.

4.1.1 General Features

The optimum hopper load is that which fills half the volume of the hopper, and the delivery tube should rise just above the maximum level of parts in the hopper. The inside silhouette of the delivery tube must be designed to accept only correctly oriented parts one at a time. The linear velocity of the delivery tube should be no greater than 0.6 m/s.

4.1.2 Specific Applications

Figures 4.3 through 4.5 show some specific results for the feeding of disks, cylinders, and square prisms, respectively. From these figures it is possible to estimate the feed rates obtainable for various parts.

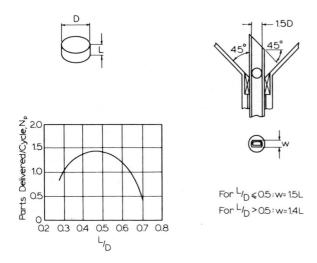

Fig. 4.3 Performance of reciprocating tube hopper when feeding disk-shaped parts.

Fig. 4.4 Performance of reciprocating tube hopper when feeding cylindrical parts.

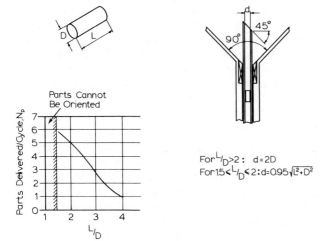

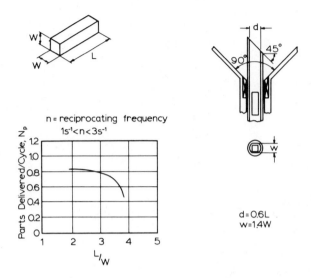

Fig. 4.5 Performance of reciprocating tube hopper when feeding square prisms.

4.2 Centerboard Hopper Feeder

Figure 4.6 shows a typical centerboard hopper feeder. Basically, this consists of a hopper in which the parts are placed at random and a blade with a shaped track along its upper edge which is periodically pushed upward through the mass of parts. The blade will thus catch a few parts on its track during each cycle, and when the blade is in its highest position (as shown in the figure), it is aligned with a chute and the parts will slide down the track and into the chute. The centerboard hopper illustrated is suitable for feeding cylindrical parts.

4.2.1 Maximum Track Inclination

One of the important parameters in a centerboard hopper design is the angle of inclination of the track when the blade is in its highest position (θ_m in Fig. 4.6). It is assumed for the purposes of the following analysis that the cam drive is arranged so that the blade is lifted rapidly to its highest position, allowed to dwell for a period while the parts slide into the chute, and then rapidly returned to its lowest position when the track is horizontal and aligned with the bottom of the hopper.

Clearly, there is a limit on the deceleration of the blade on its upward stroke; otherwise, the parts leave the track and are thrown

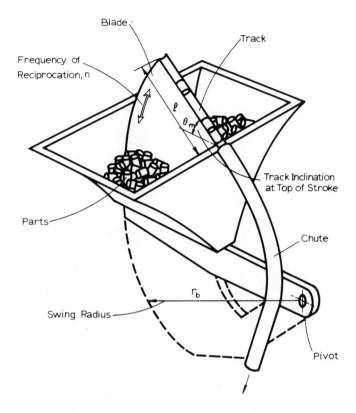

Fig. 4.6 Centerboard hopper feeder.

clear of the feeder. Thus, for a given deceleration, an increase in the angle θ_m increases the time taken for the blade to complete its upward motion. However, with larger values of θ_m the time taken for the parts to slide off the track is less and in choosing θ_m to give maximum frequency of reciprocation and hence maximum feed rate, a compromise must be sought.

The tendency for a part to leave the track during the upward motion of the blade is greatest at the end of the track farthest away from the pivot. The forces acting on a part in this position are shown in Fig. 4.7 and, from the figure, the condition for the reaction between the part and the track to become zero is given by

$$m_p \ddot{\theta}(r_b - \frac{L}{2}) = -m_p g \cos \theta_m \qquad (4.9)$$

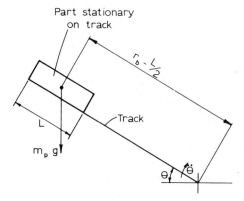

Fig. 4.7 Forces acting on part during upward motion of blade.

where L is the length of part, m_p the mass of part, r_b the radius from pivot to upper end of track, θ_m the maximum angle between track and horizontal, and $\ddot{\theta}$ the angular acceleration of track.

Thus, the maximum angular deceleration of the blade is approximately given by

$$-\ddot{\theta} = \frac{g \cos \theta_m}{r_b} \qquad (4.10)$$

if L is small compared with r_b.

For simplicity, it is now assumed that the drive to the blade is designed to give, during the period of the upward motion of the blade, (1) a constant acceleration of $(g \cos \theta_m)/r_b$ followed by (2) a constant deceleration of $(g \cos \theta_m)/r_b$. Under these conditions, the total time t_1 taken to lift the blade so that the track is inclined at an angle θ_m to the horizontal is given by

$$t_1^2 = \frac{4r_b \theta_m}{g \cos \theta_m} \qquad (4.11)$$

It is now assumed that when the blade is in its highest position, it dwells for a period t_2, just sufficient to allow the parts to slide down the track. This is given, in the worst case, by the time taken for one part to slide the whole length of the track. The forces acting on a part under these circumstances are shown in Fig. 4.8 and resolving in a direction parallel to the track gives

$$m_p a = m_p g \sin \theta_m - \mu_d m_p g \cos \theta_m \qquad (4.12)$$

where a is the linear acceleration of the part down the track and μ_d is the coefficient of dynamic friction between the part and the track. The minimum dwell period t_2 is now given by

$$t_2^2 = \frac{2\ell}{g(\sin \theta_m - \mu_d \cos \theta_m)} \tag{4.13}$$

where ℓ is the total length of the track.

If the time taken to return the blade to its lowest position is now assumed to be the same as the time for the up stroke, then the total period t_f of the feeder cycle is given by

$$t_f = 2t_1 + t_2 = 2\left(\frac{4r_b\theta_m}{g \cos \theta_m}\right)^{1/2} + \left[\frac{2\ell}{g(\sin \theta_m - \mu_d \cos \theta_m)}\right]^{1/2} \tag{4.14}$$

Equation (4.14) consists of two terms; one that increases as θ_m is increased and one that decreases as θ_m is increased. An optimum value of θ_m always exists that gives the minimum period t_f and hence a maximum theoretical feed rate. It can be shown mathematically that this optimum value of θ_m is a function only of μ_d and the ratio r_b/ℓ. However, the resulting expression is unmanageable, but the curve shown in Fig. 4.9a gives the solution for a practical value of r_b/ℓ of 2.0. For example, with a coefficient of dynamic friction of 0.4, the optimum track angle would be approximately 36 degrees.

Figure 4.9b shows how the maximum frequency of the blade cycle n_{max} (given by $1/t_f$) varies as the coefficient of friction between part and track is changed and when the ratio r_b/ℓ is 2.0. It can be seen that for large values of μ_d in the range 0.4 to 0.8, the maximum blade frequency varies by only 10 or 15%. The maximum blade frequency is more sensitive to changes in the length ℓ of the track, and for longer tracks the frequency is lower. However, it should be remem-

Fig. 4.8 Forces acting on part as it slides down track.

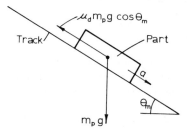

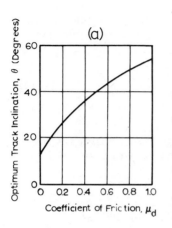
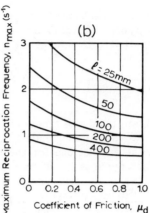

Fig. 4.9 Characteristics of centerboard hopper ($r_b / \ell = 2.0$).

bered that for a given size of part, a longer track on average picks up a greater number of parts per cycle, and hence the mean feed rate may increase.

The maximum number of parts that may be selected during each cycle is given by ℓ/L. In practice the average number selected is less than this and if E is taken to be the efficiency of a particular design, the average number of parts fed during each cycle is given by $E\ell/L$ and the mean feed rate F of the hopper feeder is given by

$$F = \frac{nE\ell}{L} \qquad (4.15)$$

where the blade frequency n is given by

$$n = \frac{1}{t_f} \qquad (4.16)$$

In practice, the values of the efficiency E must be obtained from experiments.

4.2.2 Load Sensitivity and Efficiency

For a centerboard hopper feeder working at a constant frequency any variation in feed rate as the hopper gradually empties will be due to changes in the efficiency E. This has been defined as the ratio between the average number of parts selected during one cycle and the maximum number that can be selected. Figures 4.10 and 4.11 show

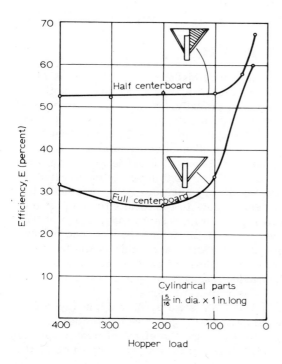

Fig. 4.10 Load sensitivity of centerboard hopper.

Fig. 4.11 Frequency of part selection of centerboard hopper.

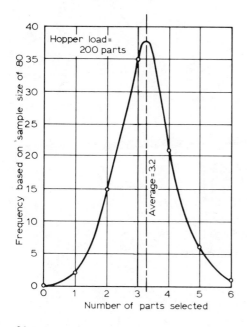

the results of tests on an experimental feeder where $\ell/L = 6$, $r_b/\ell = 2$, and $\theta_m = 54$ degrees. The lower curve in Fig. 4.10 shows how the efficiency E varied as the hopper gradually emptied, and it is interesting to note that a rapid increase in E occurs when less than 100 parts remain in the hopper. The upper curve in Fig. 4.10 shows that this high efficiency can be maintained for all hopper loadings if a baffle is placed on one side of the hopper blade. The baffle would therefore appear to affect the orientation of those parts likely to be selected by the blade. Clearly, the load sensitivity characteristics obtained with this latter design approach very closely to the ideal situation. Each of the experimental points in Fig. 4.10 represents the average of 80 results. In Fig. 4.11 one set of 80 results (for a hopper load of 200 parts) is plotted in the form of a histogram and exhibits the familiar normal distribution. The average number of parts selected in this case was 3.2, and since the maximum number of parts that could be selected by the blade was 6, this represents an efficiency of 0.53.

Figures 4.12 through 4.14 illustrate some specific applications for the centerboard hopper feeder: the feeding of disks, cylinders, and headed parts, respectively. In each case the effect of part proportions on the efficiency is presented. These graphs allow an estimate of the feed rate to be made.

Fig. 4.12 Performance of centerboard hopper when feeding disks (ℓ, length of track).

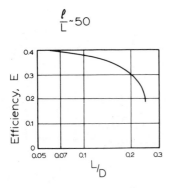

Enlarged section
through track

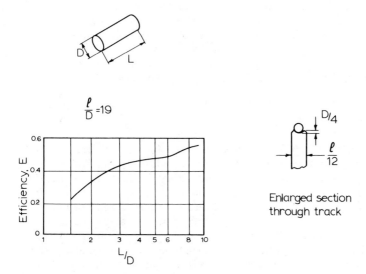

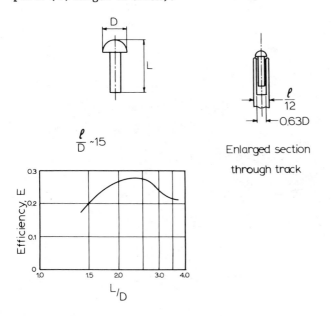

Fig. 4.13 Performance of centerboard hopper when feeding cylinders (ℓ , length of track).

Fig. 4.14 Performance of centerboard hopper when feeding headed parts (ℓ , length of track).

A final design consideration is the inclination of the sloping sides of the hopper. If the inclination is too great, there is a possibility that parts will jam between the hopper wall and the blade when the blade is moving downward. This situation is identical to that of the reciprocating tube hopper, and the analysis indicated that the included angle between the hopper wall and the blade should be 60 degrees.

4.3 Reciprocating Fork Hopper Feeder

A reciprocating fork hopper is shown in Fig. 4.15 and is suitable only for feeding headed parts. It consists of a shallow cylindrical bowl that rotates about an axis inclined at approximately 10 degrees to the vertical and a fork that reciprocates in the vertical plane about point A. In its lowest position, the fork is horizontal and rotation of the bowl causes parts to be caught in the fork. The fork then lifts a few parts by their heads to a height sufficient to cause the parts to slide off the fork and into the delivery chute. The analysis for the maximum fork inclination and the maximum rate of reciprocation would be similar to that presented above for a centerboard hopper. The number of parts selected by the fork per cycle would be obtained by experiment.

Fig. 4.15 Reciprocating fork hopper.

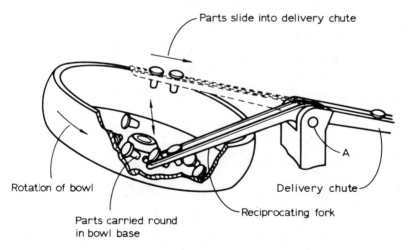

4.4 External Gate Hopper Feeder

An external gate hopper basically consists (Fig. 4.16) of a rotating cylinder having slots in its wall where the cylindrical parts, if oriented correctly, can nest against the wall of the stationary outer sleeve. At some point, as the cylinder rotates, the slots pass over a gate in the outer sleeve which allows the parts to drop one by one into the delivery chute. The tumbling action caused by rotation of the cylinder provides repeated opportunities for parts to fall into the slots and subsequently to pass through the external gate into the chute.

4.4.1 Feed Rate

Figure 4.17a shows an enlarged cross section of the slot and part just before the part falls through the gate. In the following analysis, an equation is developed for the maximum peripheral velocity of the inner cylinder for feeding to occur. Clearly, if the velocity is too high, the part will pass over the gate. At some limiting velocity v the part will neither fall through the gate nor pass over but will become jammed between corners B and C of the slot and gate as shown in Fig. 4.17b. With any velocity below v the part will drop through the gate as shown in Fig. 4.17c. The position shown in Fig. 4.17a represents the point at which the part starts to fall. In Fig. 4.17b the part has moved from this position a horizontal distance

$$D - 0.5(D^2 - h_g^2)^{1/2}$$

which, at a velocity v represents a time interval of

$$\frac{D - 0.5(D^2 - h_g^2)^{1/2}}{v}$$

During this time the part has fallen a distance $(D/2 - h_g/2)$ and if it is assumed that the part has fallen freely, the time taken is given by $[2(D/2 - h_g/2)/g]^{1/2}$. Thus by equating these times, the limiting velocity is given by

$$\frac{D - 0.5(D^2 - h_g^2)^{1/2}}{v} = \left(\frac{D - h_g}{g}\right)^{1/2} \tag{4.17}$$

To give the largest values of v, the gap h_g between the cylinder and sleeve should be as large as possible. For values of h_g greater than $D/2$ there is a danger of the parts becoming jammed between the corner B of the slot and the inner surface of the sleeve. Thus, taking $h_g = D/2$, Eq. (4.17) becomes, after rearrangement,

$$v = 0.802(Dg)^{1/2} \tag{4.18}$$

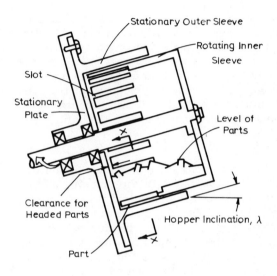

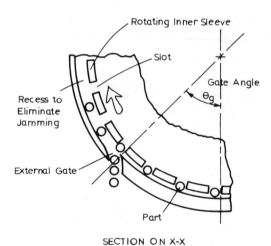

SECTION ON X-X

Fig. 4.16 External gate hopper. General data: maximum peripheral velocity, 0.5 m/s; gate angle θ_g, 0.79 to 1.13 rad (45 to 65 degrees); hopper inclination λ, 0.17 to 0.26 rad (10 to 15 degrees).

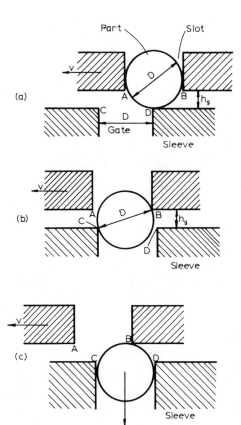

Fig. 4.17 Various stages in motion of part when passing through the gate of an external gate hopper.

If a_s is now taken to be the centerline distance between adjacent slots of the cylinder, the maximum feed rate F_{max} from the feeder is

$$F_{max} = \frac{v}{a_s} = 0.802(Dg)^{1/2} \qquad (4.19)$$

In general, only a proportion of the slots catches parts and if E is taken to be the efficiency of the feeder, the actual feed rate is given by

$$F = \frac{0.802(Dg)^{1/2}E}{a_s} \qquad (4.20)$$

The type of feeder analyzed above is generally used for feeding rivets where the slots in the inner cylinder are open ended to allow for the rivet heads. If the diameter of the rivet head is D_h, the minimum theoretical distance between the centers of the slots is D_h and Eq. (4.20) becomes

$$F = \frac{0.802(Dg)^{1/2}E}{D_h} = 0.802E\left(\frac{D}{D_h}\right)\left(\frac{g}{D}\right)^{1/2} \qquad (4.21)$$

and for plain cylindrical parts

$$F = 0.802E\left(\frac{g}{D}\right)^{1/2} \qquad (4.22)$$

where the partition between parts is very small.

This analysis has considered the maximum possible feed rate from an external gate feeder. In practical designs, the actual feed rate will be less than this because of mechanical limitations, but the result indicates that certain trends might be expected from this type of feeder, and these are now summarized:

1. The maximum unrestricted feed rate is inversely proportional to the square root of the diameter of a cylindrical part.
2. When feeding rivets with the same head diameter the maximum feed rate is proportional to the square root of the shank diameter; when feeding rivets where the ratio of shank diameter to head diameter is fixed, the maximum feed rate is inversely proportional to the square root of the shank diameter.
3. If a high feed rate is required, the slots in the inner cylinder of the feeder should be as close as possible.

4.4.2 Load Sensitivity and Efficiency

The unrestricted feed rate for a given design of feeder depends on E, the efficiency. This may be affected by the load in the hopper, the angle of inclination of the feeder axis, and the position of the external gate. Tests have shown that the most significant of these variables is the angle of inclination λ of the feeder axis. The results presented in Fig. 4.18 indicate that λ should be as low as possible for maximum efficiency.

However, it should be realized that a practical limitation exists because, as λ is reduced, the capacity of the hopper is also reduced and a compromise must therefore be reached in any given design. A typical figure for λ is between 10 and 15 degrees. The results in Fig. 4.18 also show that the efficiency of the feeder increases rapidly

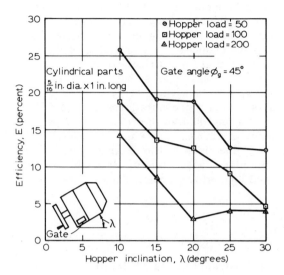

Fig. 4.18 Effect of hopper inclination on efficiency of external gate hopper.

Fig. 4.19 Effect of gate angle on efficiency of external gate hopper.

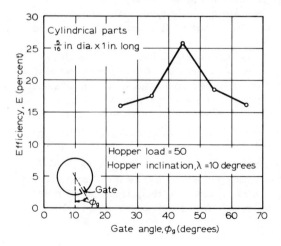

as the hopper empties. Figure 4.19 shows the effect of the angular position ϕ_g of the external gate on the efficiency. It is clear that for values of ϕ_g greater than 90 degrees, the efficiency would become zero, and for very small values of ϕ_g the chances of parts falling into the slots is reduced. The results show that an optimum exists when ϕ_g is approximately 45 degrees, and Fig. 4.18 shows that, with this optimum value and with the lowest practical value of λ, 10 degrees, the minimum efficiency was 0.14. This represents [from Eq. (4.20) when $a_s = 2D$] a maximum possible feed rate of approximately 2 parts per second with cylindrical parts of 12.3 mm (5/16 in.) diameter.

Figures 4.20 and 4.21 show applications of the external gate hopper to the feeding of rivets and cylinders, respectively. In each case the feed rate per slot FR_s is plotted against the peripheral velocity of the sleeve v.

4.5 Rotary Disk Feeder

A typical rotary disk feeder is illustrated in Fig. 4.22. This feeder consists of a disk with a number of slots machined radially in its face and mounted at a steep angle to the horizontal, so that it forms the base of a stationary hopper. As the disk rotates, the parts in the hopper are disturbed by the ledges next to the slots. Some parts are caught in the slots and carried around until, when each slot in turn reaches the highest position, it becomes aligned with a delivery chute down which the parts slide. A stationary circular plate at the center of the disk prevents the parts sliding out of the slots until they are aligned with the chute.

In some designs of rotary disk feeders, the length of the slots will allow more than one part per slot to be selected during each revolution of the disk. This design will be analyzed first and it will be assumed that, to give the greatest efficiency, the disk is indexed with sufficient dwell to allow all the parts selected in each slot to slide down the chute.

4.5.1 Indexing Rotary Disk Feeder

If a Geneva mechanism is employed to index a rotary disk feeder, the time for index will be approximately equal to the dwell period. For the design illustrated in Fig. 4.22, the time t_s required for all parts in one slot to slide into the delivery chute is given by Eq. (4.13), which is

$$t_s^2 = \frac{2\ell}{g(\sin\theta - \mu_d \cos\theta)} \qquad (4.23)$$

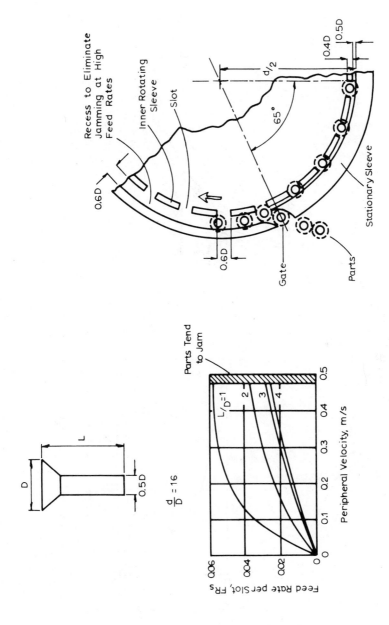

Fig. 4.20 Performance of external gate hopper when feeding headed parts.

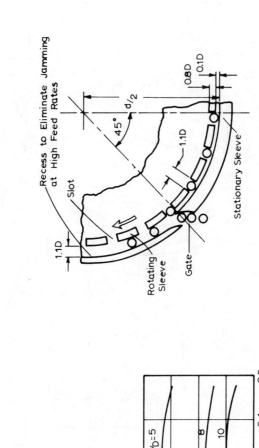

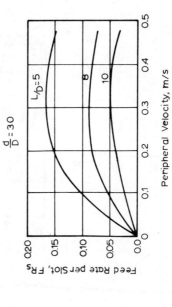

Fig. 4.21 Performance of external gate hopper when feeding cylinders.

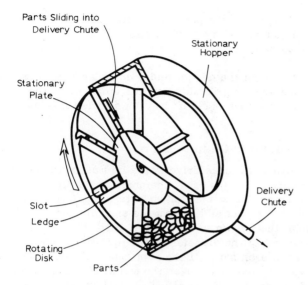

Fig. 4.22 Rotary disk feeder.

where ℓ is the length of the slot, θ the inclination of the delivery chute, and μ_d the coefficient of dynamic friction between the part and the chute.

With a Geneva drive, the total period of an indexing cycle t_i is therefore given by

$$t_i = 2t_s = \left[\frac{8\ell}{g(\sin \theta - \mu_d \cos \theta)} \right]^{1/2} \qquad (4.24)$$

If L is the length of a part, the maximum number that may be selected in a slot is ℓ/L. However, in practice the average number selected will be less than this. If E is taken to be the efficiency of the feeder, the feed rate F will be given by

$$F = \frac{E\ell}{Lt_i} = E \left[\frac{\ell g(\sin \theta - \mu_d \cos \theta)}{8L^2} \right]^{1/2} \qquad (4.25)$$

It can be seen from Eq. (4.25) that if E is assumed to remain constant:

1. The feed rate is independent of the number of slots in the disk.
2. For a given feeder the feed rate is inversely proportional to the length of the part.

3. For maximum feed rate with a given part, μ_d should be as low as possible and both the delivery chute angle θ and the slot length ℓ should be as large as possible.

It is clear, however, that with the design under consideration, the feed rate will reduce as the hopper gradually empties until, when the hopper is almost empty no more than one part may be selected in each slot.

4.5.2. Rotary Disk Feeder with Continuous Drive

A rotary disk feeder with continuous drive would be most suitable for feeding disk-shaped parts. In this case, the analysis for the maximum feed rate would be similar to that for an external gate hopper because the situation where the part slides from the slot into the delivery chute is similar to that shown in Fig. 4.17. With this device the slot length would be equal to the diameter of the part D and only one part could be selected in each slot. If the rotational speed of the disk were too high, the parts would pass over the mouth of the delivery chute and feeding would not occur. If the effect of friction is considered negligible because of the large angle of inclination of the disk, the feed rate at maximum rotational speed will be given by Eq. (4.22), which is

$$F = 0.802E\left(\frac{g}{D}\right)^{1/2} \tag{4.26}$$

where E is the efficiency of feeder (that is, the average number of parts selected per cycle divided by the number of slots.

The foregoing analyses have considered the theoretical maximum feed rate from a rotary disk feeder, both with indexing drive and with continuous drive. The results indicate that the following trends would be expected from this type of feeder:

1. For an indexing rotary disk feeder with long slots, the maximum feed rate is inversely proportional to the length of the part and proportional to the square root of the slot length. For high feed rates, the slope of the delivery chute should be as large as possible and the coefficient of friction between the part and the chute should be as low as possible.
2. For a feeder with continuous drive the maximum feed rate for disk-shaped parts is inversely proportional to the square root of the diameter of the part.

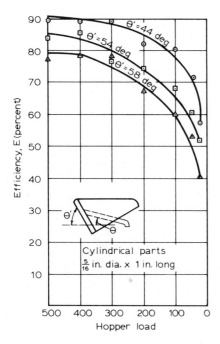

Fig. 4.23 Load sensitivity of rotary disk feeder.

4.5.3 Load Sensitivity and Efficiency

Tests were conducted on an indexing rotary disk feeder with eight slots, each able to carry two cylindrical parts, 25 mm (1 in.) in length and 8 mm (5/16 in.) in diameter. The results are presented in Fig. 4.23, which shows that, as would be expected, the efficiency reduces as the hopper empties. This is because for small loads, the mass of parts only partly covers the slots and at best only one part per slot can be selected during each cycle. The figure shows that both the efficiency and the load sensitivity characteristics are improved as the angle of inclination θ' of the disk is reduced. Unfortunately, this also reduces the inclination θ of the delivery chute and increases the time taken for the parts to slide out of the slots. Clearly, in any given design a compromise is necessary. In practice, typical values for the angle of inclination of the disk are from 55 to 69 degrees and for the delivery chute, an inclination of 35 degrees is typical. Figures 4.24 and 4.25 show applications of the rotary disk feeder to the feeding of disks and cylinders, respectively.

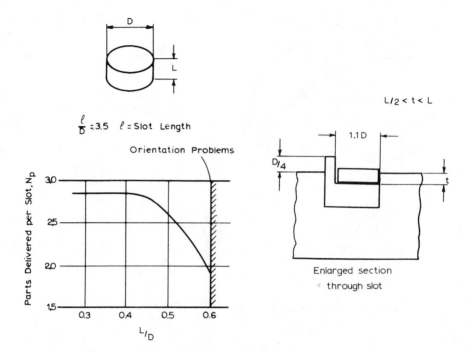

Fig. 4.24 Performance of rotary disk feeder when feeding disks.

4.6 Centrifugal Hopper Feeder

The centrifugal hopper feeder shown in Fig. 4.26 is particularly suitable for feeding plain cylindrical parts. In this device, the parts are
placed in a shallow cylindrical hopper whose base rotates at constant
speed. A delivery chute is arranged tangentially to the stationary
wall of the hopper, and parts adjacent to this wall which have become
correctly oriented, due to the general circulation, pass into the
delivery chute. No orienting devices are provided in the hopper and
therefore parts must be taken off in the attitude which they naturally
adopt in the hopper, as indicated in the figure.

4.6.1 Feed Rate

If a part is moving with constant velocity v around the inside wall
of a centrifugal hopper, the radial reaction at the hopper wall is equal
to the centrifugal force $2m_p v^2/d$, where m_p is the mass of the part
and d the diameter of the hopper. The frictional force F_w at the
hopper wall tends to resist the motion of the part and is given by

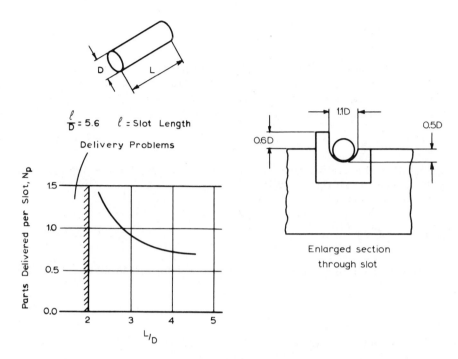

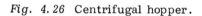

Fig. 4.25 Performance of rotary disk feeder when feeding cylinders.

Fig. 4.26 Centrifugal hopper.

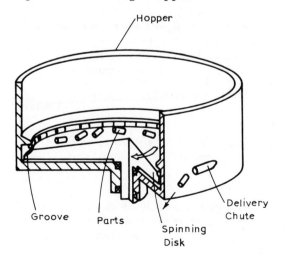

$$F_w = \frac{2\mu_w m_p v^2}{d} \qquad (4.27)$$

where μ_w is the coefficient of friction between the part and the hopper wall. When the peripheral velocity of the spinning disk is greater than v, the disk slips under the part and the frictional force F_b between the part and the spinning disk is given by

$$F_b = \mu_b m_p g \qquad (4.28)$$

where μ_b is the coefficient of friction between the part and the spinning disk. Since under this condition $F_b = F_w$, setting Eq. (4.27) equal to Eq. (4.28) gives

$$v = \left(\frac{g\mu_b d}{2\,\mu_w}\right)^{1/2} \qquad (4.29)$$

and the maximum feed rate F_{max} of parts of length L is given by

$$F_{max} = \frac{v}{L} = \frac{(g\mu_b d/2\mu_w)^{1/2}}{L} \qquad (4.30)$$

and the actual feed rate F may be expressed as

$$F = E\frac{(g\mu_b d/2\mu_w)^{1/2}}{L} \qquad (4.31)$$

where E is the feeder efficiency.

Equation (4.31) shows that the unrestricted feed rate from a centrifugal hopper is proportional to the square root of the hopper diameter and inversely proportional to the length of the parts.

Using Eq. (4.29), the maximum rotational frequency n_{max} of the spinning disk, above which no increase in feed rate occurs, is

$$n_{max} = \frac{v}{\pi d} = \frac{[(g/2d)(\mu_b/\mu_w)]^{1/2}}{\pi} \qquad (4.32)$$

This equation is plotted in Fig. 4.27, and can be used to choose the maximum rotational frequency of the hopper.

4.6.2 Efficiency

The overall efficiency E of the hopper can only be determined by experiment. Figure 4.28 shows results of experiments showing the effect of rotational frequency on the efficiency when feeding plain cylinders.

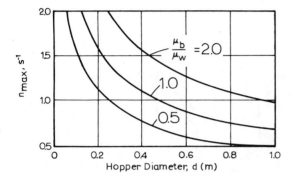

Fig. 4.27 Maximum rotational frequency for centrifugal hopper, where n_{max} is the maximum rotational frequency of the disk, μ_b the coefficient of friction between the part and the spinning disk, and μ_w the coefficient of friction between the part and the stationary hopper wall.

Fig. 4.28 Performance of centrifugal hopper when feeding cylinders (d, hopper diameter).

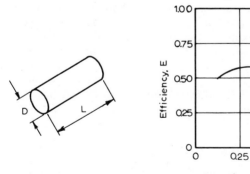
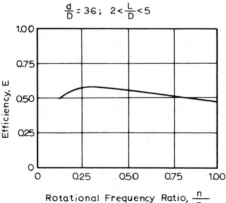

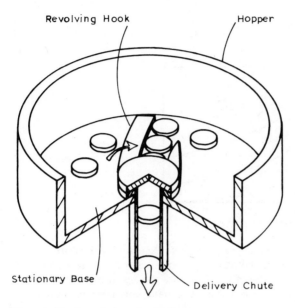

Fig. 4.29 Revolving hook hopper.

4.7 Revolving Hook Hopper Feeder

The revolving hook hopper feeder shown in Fig. 4.29 consists of a hopper with a flat base with a hole in the center which forms the beginning of the delivery chute. Revolving about the center of the base and offset so as to clear the hole is a curved wiper blade which extends to the outer edge of the hopper base. Rotation of the hook guides the parts along the leading edge of the hook, toward the hole at the center of the hopper and hence to the delivery chute.

The thickness of the revolving hook is the same as that of the parts. A wiper having a tapered outside edge and the hook form a passage whose width is less than twice the diameter of the parts. A hood is attached to the hook above the center hole so that parts do not block the delivery chute.

4.7.1 Rotational Frequency and Feed Rate

If the rotational frequency of the hook is too high, centrifugal action prevents the parts from moving inward toward the delivery chute and a drastic reduction in feed rate occurs. The effect of hopper diameter on the critical rotational frequency at which this reduction in feed rate occurs is presented in Fig. 4.30. Figure 4.31 presents experimental results for the performance of the feeder when feeding disk-shaped parts.

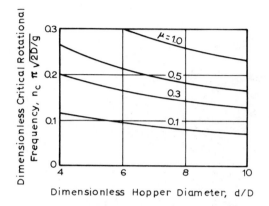

Fig. 4. 30 Critical rotational frequencies for revolving hook hopper. n_c is the critical rotational frequency, d the hopper diameter, D the part diameter, μ the coefficient of friction between the part and the stationary base, and g = 9810 mm/s^2. Example: If D = 30 mm, d = 240 mm, and μ = 0.4, then d/D = 8 and from the graph $n_c \pi (2D/g)^{1/2}$ = 0.17. The critical rotational frequency n_c is 0.69 s^{-1} (42 rpm).

Fig. 4. 31 Performance of revolving hook hopper when feeding disks.

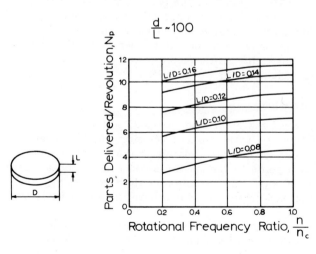

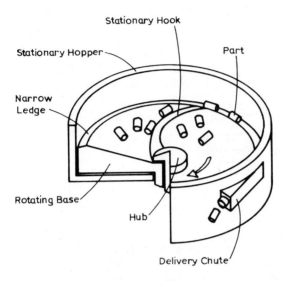

Fig. 4.32 Stationary hook hopper.

4.8 Stationary Hook Hopper Feeder

This parts feeder (Fig. 4.32) operates on exactly the same principle
as the revolving hook hopper feeder, the only difference being that,
in this case, the hook is stationary and the base of the hopper rotates
slowly. The parts are guided along the edge of the hook toward the
periphery of the hopper, where they are eventually deflected into the
delivery chute by a deflector mounted on the hopper wall. One advantage of this type of feeder is its gentle feeding action, and this makes
it suitable for feeding delicate parts at low speed.

4.8.1 Design of the Hook

Ideally, the stationary hook should be designed so that a part
travels along its leading edge with constant velocity. With reference
to Fig. 4.33a, the velocity of the part relative to the hook is designated
v_{12} and the velocity of a point on the moving surface of the base relative
to the hook is designated v_{32}. The direction of the velocity, v_{13}, of
the part relative to the base is then obtained from the velocity diagram,
Fig. 4.33b.

The frictional force between the part and the moving surface is in
the opposite direction to the relative velocity v_{13} and is the force that
causes the part to move along the hook. The direction of this force
is specified by ψ in Fig. 4.33b. By the law of sines

$$\frac{v_{12}}{v_{32}} = \frac{\sin \psi}{\sin (\theta + \psi)} \qquad (4.33)$$

where θ is the angle between the velocity vector v_{32} and the tangent to the hook at the point under consideration.

Neglecting the inertia force, the external forces acting on the part in the horizontal plane are shown in Fig. 4.33c. Resolving these forces parallel and normal to the hook tangent, the conditions for equilibrium are

$$\mu_r N_r = \mu_p W \cos (\theta + \psi)$$
$$N_r = \mu_p W \sin (\theta + \psi) \qquad (4.34)$$

where μ_r is the coefficient of dynamic friction between the part and the hook, μ_p the coefficient of sliding friction between the part and the moving surface, N_r the normal reaction between the part and the hook, and W the weight of the part. Combining Eqs. (4.34) gives

$$\mu_r = \cot (\theta + \psi) \qquad (4.35)$$

Fig. 4.33 Analysis of stationary hook hopper.

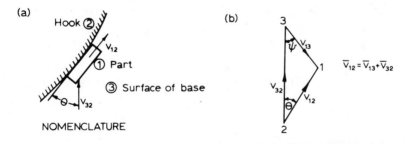

(a) Hook ② ... v_{12} ① Part ③ Surface of base — NOMENCLATURE

(b) VELOCITY DIAGRAM $\overline{V}_{12} = \overline{V}_{13} + \overline{V}_{32}$

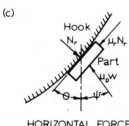

(c) HORIZONTAL FORCES ACTING ON PART

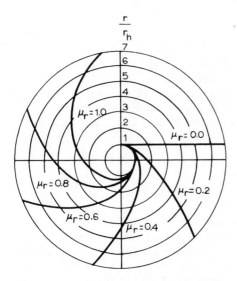

Fig. 4.34 Hook shapes for stationary hook hopper, where r_h is the radius of the hub, r is the radial location of a particular point on the hook. Maximum peripheral velocity is 0.4 m/s.

Equations (4.33) and (4.35) can be combined to eliminate ψ:

$$\frac{v_{12}}{v_{32}} = \cos\theta - \mu_r \sin\theta \qquad (4.36)$$

The velocity of a point on the base relative to the hook is given by

$$v_{32} = 2\pi rn \qquad (4.37)$$

where r is the distance from the point under consideration to the center of the hopper and n the rotational frequency of the hopper base. For constant speed of the part along the hook, Eqs. (4.36) and (4.37) can be combined to give

$$r(\cos\theta - \mu_r \sin\theta) = K \qquad (4.38)$$

where K is a constant given by $v_{12}/2\pi n$.

An explicit equation defining the shape of the hook is not readily obtainable from Eq. (4.38). However, a numerical method was used to develop the hook shapes giving constant-speed feeding for various values of the coefficient of sliding friction between the part and the

hook; these hook shapes are presented in Fig. 4.34. A computer
program constructed the hook shapes using straight-line segments
starting at the hub and working outward. The results in Fig. 4.34
are plotted in a dimensionless form in which the radius to a point on
the hook is divided by the hub radius.

In using these results to design the shape of a hook for a particular
application, it would always be preferable to use a larger than expected
value of the coefficient of sliding friction between the part and the
hook. This would ensure that parts would not decelerate as they
moved along the hook and thus would tend to avoid jamming in the
feeding process.

4.8.2 Feed Rate

It is apparent that the velocity of the parts along the hook will
determine the maximum feeder output. For parts of a given length
being fed end to end, the maximum output per unit of time is determined
by dividing the velocity at the hub by the individual part length.
Figure 4.35 shows how the maximum feed rate from the feeder varies
with the length of parts for various rotational frequencies.

Fig. 4.35 Maximum feed rate for stationary hook hopper, where $F = (r_h/L)(2\pi n)$, with L the length of the part and r_h the feeder hub radius.

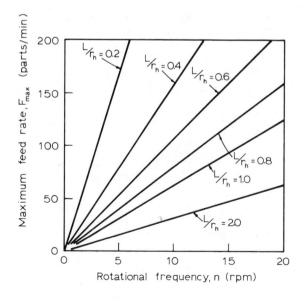

For particular applications, the actual feed rate will be lower than that given in Fig. 4.35. For a particular design of feeder this feed rate can be expressed by

$$F = \frac{E \pi d n}{L}$$

(4.39)

where d is the hopper diameter, n the rotational frequency of the disk, L the part length, and E the efficiency of the feeder. Figures 4.36 and 4.37 illustrate examples of the application of the stationary hook hopper to the feeding of cylinders and headed parts, respectively. In each graph, the effect of part proportions on the feeder efficiency is presented.

4.9 Bladed Wheel Hopper Feeder

In the bladed wheel hopper feeder (Fig. 4.38), the tips of the blades of a vertical multibladed wheel run slightly above a groove in the bottom of the hopper. The groove has dimensions such that the parts in the hopper may be accepted by the groove in one particular orientation only. Rotation of the wheel agitates the parts in the hopper and causes parts arriving at the delivery point in the wrong attitude to be pushed back into the mass of parts.

Fig. 4.36 Performance of stationary hook hopper when feeding cylinders, with load ratio $R_L = LN/\pi D$ and d/D = 36.

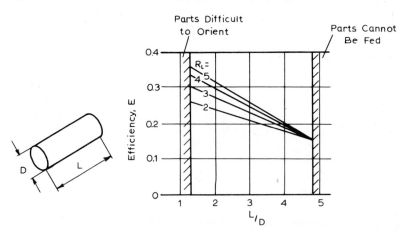

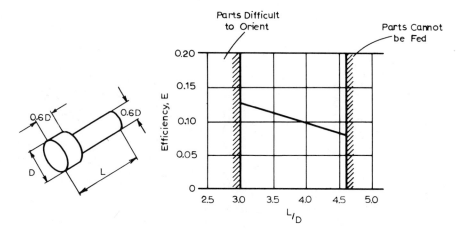

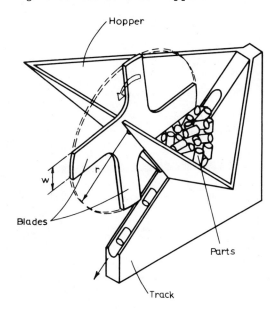

Fig. 4.37 Performance of stationary hook hopper when feeding headed
parts, with $d/D = 36$ and $0.1 < n < 0.3 \ s^{-1}$.

Fig. 4.38 Bladed wheel hopper.

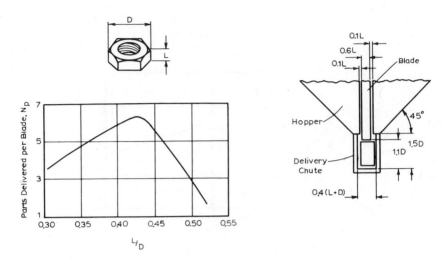

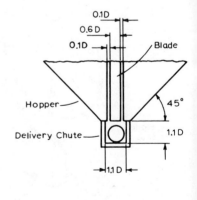

Fig. 4.39 Performance of bladed wheel hopper when feeding nuts.

Fig. 4.40 Performance of bladed wheel hopper when feeding cylinders.

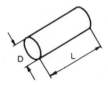

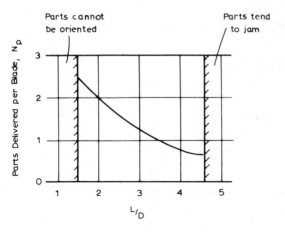

The angle of inclination of the track is generally about 45 degrees and the maximum linear velocity of the blade tip is around 0.7 m/s. Figures 4.39 and 4.40 show applications of the bladed wheel hopper to the feeding of nuts and cylinders, respectively.

4.10 Tumbling Barrel Hopper Feeder

In this feeder (Fig. 4.41), the cylindrical container, which has internal radial fins, rotates about a horizontal vibratory feed track. Parts placed in bulk in the hopper are carried upward by the fins until at some point they slide off the fin and cascade onto the vibratory feed track. The feed track is shaped to suit the required orientation of the part being fed and only retains and feeds those parts in this orientation. This feeder is suitable for feeding a wide variety of parts such as cylinders, U-shaped parts, angled parts, and prisms.

The optimum hopper load is that which fills the barrel to a height of one-fourth the inside diameter. For parts that can stack one upon another, it is necessary to provide the outlet with a stationary gate which outlines the shape of a part on the rail so that parts come out

Fig. 4.41 Tumbling barrel hopper.

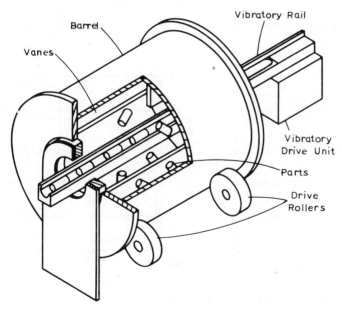

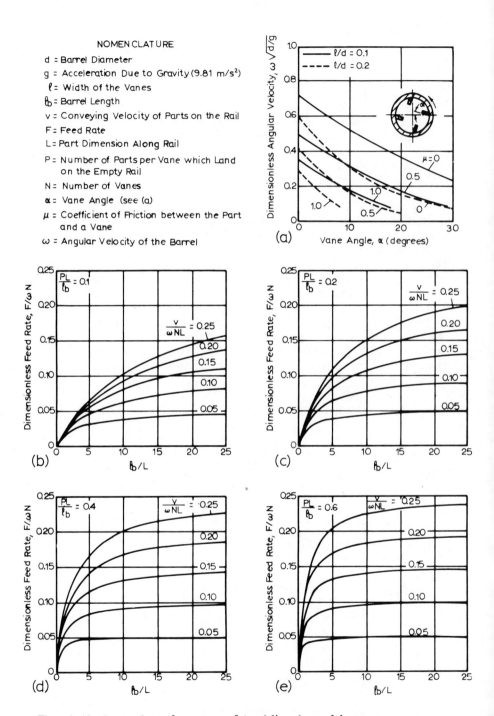

NOMENCLATURE

d = Barrel Diameter

g = Acceleration Due to Gravity (9.81 m/s²)

ℓ = Width of the Vanes

ℓ_b = Barrel Length

v = Conveying Velocity of Parts on the Rail

F = Feed Rate

L = Part Dimension Along Rail

P = Number of Parts per Vane which Land on the Empty Rail

N = Number of Vanes

α = Vane Angle (see (a))

μ = Coefficient of Friction between the Part and a Vane

ω = Angular Velocity of the Barrel

Fig. 4. 42 General performance of tumbling barrel hopper.

90

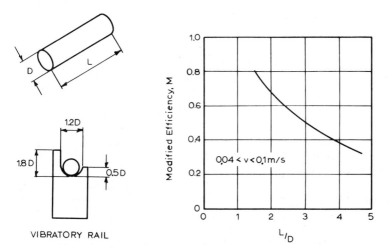

VIBRATORY RAIL

Fig. 4.43 Performance of tumbling barrel hopper when feeding cylinders. The feed rate is Mv/L + 0.7 part per second, where M is the modified efficiency, v the conveying velocity on the vibratory rail (m/s), and L the part length (m). The ratio of barrel length to diameter d is 1.2; d/D = 21.

Fig. 4.44 Performance of tumbling barrel hopper when feeding U-shaped parts. The feed rate is Ev/L, where E is the efficiency, v the conveying velocity on the vibratory rail, and L the part length. The ratio of barrel length to diameter d is 1.2; d/W = 13; and 0.4 < L/W < 1.2.

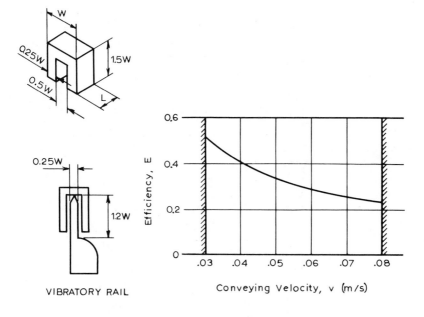

VIBRATORY RAIL

in a single layer only. Rubber or cork may be placed along the inside wall of the barrel to reduce noise and part damage.

4.10.1 Feed Rate

For a part where motion relative to the vanes is one of pure translation (that is, the part does not roll), the rotational frequency which maximizes the number of parts that land on a rail passing through the barrel center can be determined from Fig. 4.42a. The result may not be valid for parts whose motion relative to the vanes contains rotational components.

Figure 4.42b through e can be used to estimate the values of the barrel design parameters that achieve a given feed rate. To use the graphs it is first necessary to determine, by observation, the average number of parts that land on an empty rail per vane P.

Figures 4.43 through 4.45 show applications of the tumbling barrel hopper to the feeding of cylinders, U-shaped parts, and angled parts.

Fig. 4.45 Performance of tumbling barrel hopper when feeding angled parts. The feed rate is Ev/L, and the load ratio R_L = LN/W; E is the efficiency, v the conveying velocity on the vibratory rail, L the part length, N the number of parts in the hopper, and W the part width. The ratio of barrel length to diameter d is 1.2; d/W = 6.47.

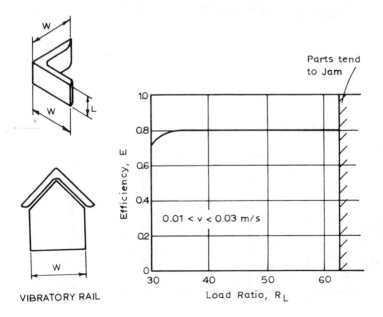

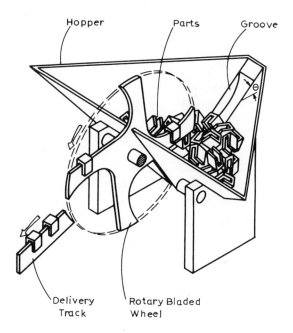

Fig. 4.46 Rotary centerboard hopper.

4.11 Rotary Centerboard Hopper Feeder

This feeder (Fig. 4.46) consists of a bladed wheel that rotates inside a suitably shaped hopper. The edges of the blades are profiled to collect parts in the desired attitude and lift them clear of the bulk of parts. Further rotation of the wheel causes the oriented parts to slide off the blade, which will then be aligned with the delivery chute. It is usual to drive the wheel intermittently by either a Geneva mechanism or a ratchet and pawl mechanism. The design of the indexing mechanism should take into account the dwell time required for a full blade to discharge all its parts when aligned with the delivery chute. A similar analysis to that used for the reciprocating centerboard hopper feeder gives the minimum values for dwell and indexing times and hence the maximum feed rate.

The peripheral velocity of the rotary bladed wheel should be no greater than 0.6 m/s. Figure 4.47 shows the application of the feeder to the feeding of U-shaped parts.

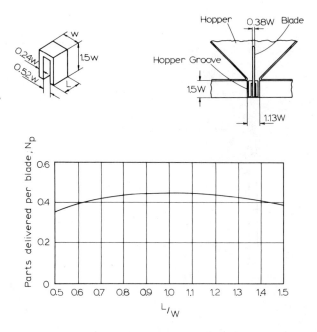

Fig. 4.47 Performance of rotary centerboard hopper when feeding U-shaped parts.

4.12 Magnetic Disk Feeder

This feeder (Fig. 4.48) consists of a container that is closed at one side by a vertical disk. The disk rotates about a horizontal axis and permanent magnets are inserted in pockets around its periphery. As the disk rotates, parts are lifted by the magnets and stripped off at a convenient point. This feeder can clearly only be used for parts of a ferromagnetic material.

The magnets should have a holding capacity of 10 to 20 times the weight of one part and have a diameter approximately equal to the major dimension of the part. The linear velocity of the magnets should be from 0.08 to 0.24 m/s. Figures 4.49 through 4.51 show applications of the feeder to the feeding of nuts, disks, and square prisms.

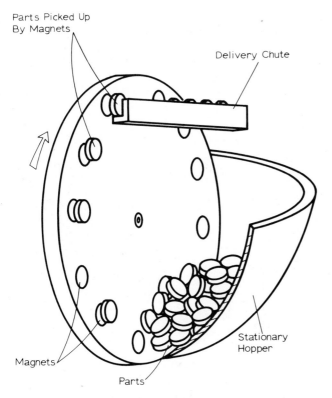

Fig. 4.48 Magnetic disk feeder.

Fig. 4.49 Performance of magnetic disk feeder when feeding nuts.

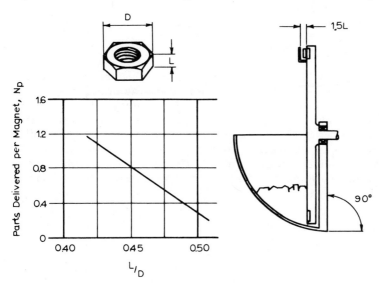

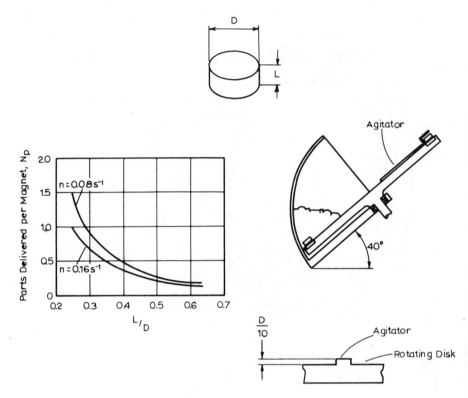

Fig. 4.50 Performance of magnetic disk feeder when feeding disks (n, rotational frequency).

4.13 Elevating Hopper Feeder

This feeder (Fig. 4.52) has a large hopper with inclined sides. Often, an agitating device is fitted to the base to encourage the parts to slide to the lowest point in the hopper. An endless conveyor belt, fitted with a series of selector ledges, is arranged to elevate parts from the lowest point in the hopper. The ledges are shaped so that they will accept parts only in the desired attitude. The parts slide off the ledges into the delivery chute, which is situated at a convenient point above the hopper. Figure 4.53 shows the application of this feeder to the feeding of cylinders.

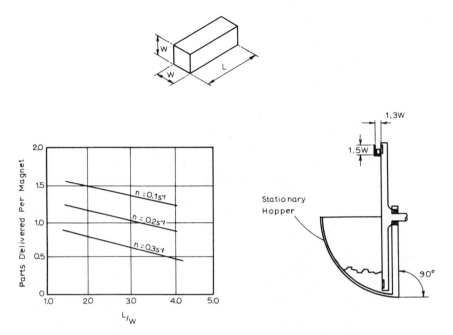

Fig. 4.51 Performance of magnetic disk feeder when feeding square prisms.

Fig. 4.52 Elevating hopper feeder.

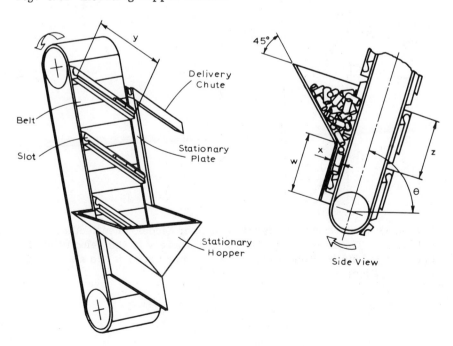

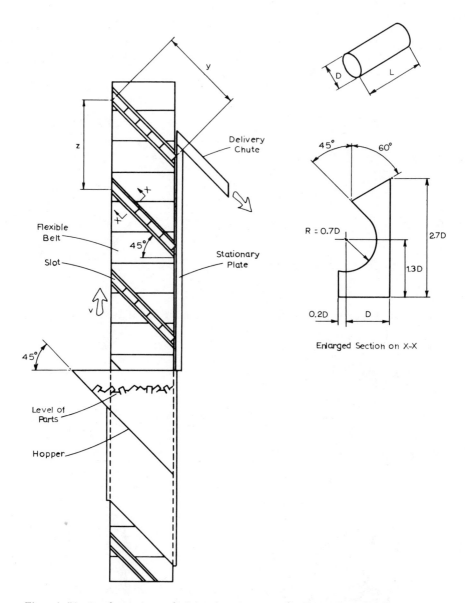

Fig. 4.53 Performance of elevating hopper feeder when feeding cylinders. The feed rate is 0.6 yv/zL, where v is the velocity of the belt, y the length of one slot, z the distance between two slots, and L the part length; L/D > 2; y/L > 4.

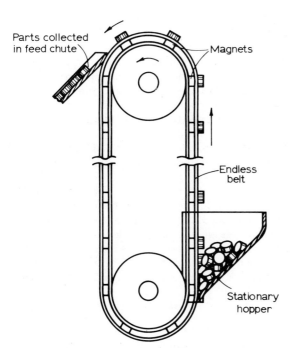

Fig. 4.54 Magnetic elevating hopper feeder.

4.14 Magnetic Elevating Hopper Feeder

The magnetic elevating hopper feeder (Fig. 4.54) is basically the
same as the elevating hopper feeder except that, instead of ledges,
permanent magnets are fitted to the endless belt. Thus, the feeder
is only suitable for handling ferromagnetic materials and cannot easily
be used for orientation purposes. With this feeder, it is usual to
strip the parts from the magnets at the top of the belt conveyor.

4.15 Magazines

An alternative means of delivering parts to an automatic assembly
machine is from a magazine. With this method, parts are stacked into
a container or magazine which constrains the parts in the desired
orientation. The magazine is then attached to the workhead of the
assembly machine. The magazines may be spring-loaded to facilitate
delivery of the parts, or alternatively, the parts may be fed under
gravity or assisted from the magazine by compressed air.

Magazines have several advantages over conventional parts feeders and some of these are described below:

1. In some cases, magazines may be designed to accept only those parts that would be accepted by the assembly machine workhead and thus can act as inspection devices. This can give a considerable reduction in the downtime on the assembly machine.
2. Magazines can often replace not only the parts feeder but also the feed track.
3. Magazines are usually very efficient feeding devices and assembly machine downtime due to feeder or feed track blockages can often be eliminated by their use.

Some benefits may be obtained if the magazines are loaded at the assembly factory. If a number of similar parts are to be used in an assembly, it may be possible to use one hopper feeder to load all the required magazines. Further, if the magazine is designed to accept only good parts, or some method of inspecting the parts is incorporated into the parts feeder, downtime will occur on the magazine loader and not on the assembly machine.

Some of the disadvantages associated with the use of magazines are as follows:

1. Magazines will generally hold considerably fewer parts than the alternative parts feeder and magazine changes must therefore be made more frequently than the refilling of the parts feeder.
2. The most suitable place to load the magazines is at the point where the part is manufactured since it is at this point that the part is already oriented. When manufacture and assembly take place in the same factory, this may not present a serious problem, but if the parts are purchased from another firm it will be much more difficult to arrange for magazine loading.
3. The cost of magazines can be considerable and prohibitive. The cost and life expectancy should be included in any economic analysis.

In some cases it is possible to use disposable magazines such as those employed for aspirin tablets. Such a magazine can be rolled up and readily fed to the workhead, which would have a suitable mechanism for removing the parts. Another alternative is where parts are blanked from strip. In this case, the final operation of separating the parts from the strip can be left to the assembly machine workhead. A further alternative often used for small blanked parts is for the blanking operation to take place on the machine just prior to the point at which the part is required in the assembly.

Chapter 5

Orientation of Parts

In an automatic assembly machine, it is necessary that the parts be
fed to the workheads correctly oriented. The devices employed to
ensure that only correctly oriented parts are fed to the workhead fall
into two groups: those that are incorporated in the parts feeder, which
are usually referred to as *in-bowl* tooling; and those that are fitted
to the chute between the feeder and the workhead, called *out-of-bowl*
tooling. The devices used for in-bowl tooling very often work on the
principle of orienting by rejection and may be termed *passive orienting
devices*. With this type of device, only those parts which, by chance,
are being fed correctly oriented pass through the device, while the
other parts fall back into the hopper or bowl. The rejected parts are
then refed and make a further attempt to pass through the orienting
devices. In some cases, devices are fitted that reorient parts. These
may be termed *active orienting devices* and although they are not as
widely applicable, they have the advantage that no reduction in feed
rate occurs due to the rejection of parts that have already been fed.
Some orienting devices are fitted between the parts feeder and the
automatic workhead. Since, with this system, rejected parts cannot
easily be returned to the parts feeder, orienting devices employed in
this way are usually of the active type.

The following chapter describes some of the more common orienting
devices and tooling employed in feeders for automatic assembly.

5.1 A Typical Orienting System

Of all the various types of feeding devices, vibratory bowl feeders
allow by far the greatest flexibility in the design of orienting devices.
Figure 5.1 shows the orienting system commonly employed to orient

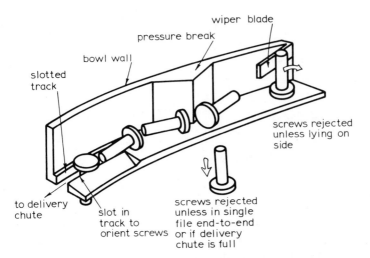

Fig. 5.1 Orientation of screws in vibratory bowl feeder.

screws in a vibratory bowl feeder. In this arrangement the first
device, a wiper blade, rejects all the screws not lying flat on the
track. The gap below the blade is adjusted so that a screw standing
on its head or a screw resting on the top of others is either deflected
back into the bowl or defected so that the screw lies flat on the track.
Clearly, the wiper blade can be applied here only if the length of the
screw is greater than the diameter of its head. The next device, a
pressure break, allows screws to pass in single file only with either
head or shank leading. Screws being fed in any other attitude will
fall off the narrow track and back into the bowl at this point. The
pressure break also performs another function; if the delivery chute
becomes full, excess parts are returned to the bottom of the bowl at
the pressure break and congestion in the chute is therefore avoided.
The last device consists of a slot in the track which is sufficiently
wide to allow the shank of the screw to fall through while retaining
the screw head. Screws arriving at the slot either with the shank
leading or with the head leading are therefore delivered with the shank
down supported by the head. In this system for orienting screws,
the first two devices are passive and the last is active.

 Although the devices described above are designed for a certain
shape of part, two of them have wide application in vibratory bowl
feeding. A pressure break is usually necessary because most feeders
are adjusted to overfeed slightly in order to ensure that the workhead
is never "starved" of parts. With this situation, unless a level-sensing
device controlling the feeder output is attached to the delivery chute,

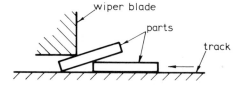

Fig. 5.2 Thin parts jammed under wiper blade.

the delivery chute is always full and a pressure break provides a
means of preventing congestion at its entrance. Second, the wiper
blade is a convenient method of rejecting parts that are resting on
top of others. In a vibratory bowl feeder this rejection often occurs
because of the pushing action of parts traveling up the track. However, care must be taken in applying the wiper blade because with
thin parts, there may be a tendency for them to jam under the blade,
as illustrated in Fig. 5.2. The tendency for this jamming to occur
will be reduced by arranging the blade so that it lies at an acute angle
to the bowl as shown in Fig. 5.1. In some cases, an alternative
approach is necessary and this is illustrated in Fig. 5.3, which shows
the orienting device commonly employed to orient washers. It can be
seen that a portion of the track is arranged to slope sideways and
down toward the center of the bowl. A small ledge is provided along
the edge of this section of the track to retain those washers that are
lying flat and in single file. Other washers will slide off the track

Fig. 5.3 Orientation of washers in vibratory bowl feeder.

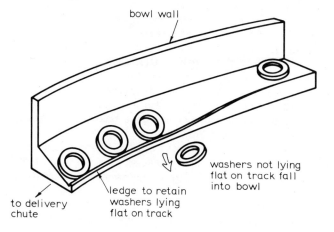

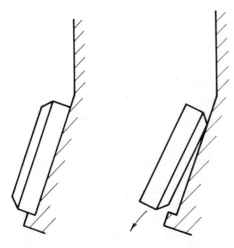

(a) Washer accepted   (b) Washer rejected

Fig. 5.4 Orientation of machined washers.

and into the bowl. With a device of this type, where the parts are
turned as they are fed, it is often necessary to arrange the design
of the track to ensure that the path of the center of gravity of the
part is not raised rapidly; otherwise, a serious reduction in feed rate
may occur.

Figure 5.4 shows a refinement made in the orienting device described
above. In this case, machined washers may be oriented by providing
a ledge sufficiently large to retain a washer being fed base down (Fig.
5.4a) but too small to retain a washer being fed base up (Fig. 5.4b).

Figure 5.5 illustrates a common type of orienting device known as
a cutout, where a portion of the track has been cut away. This device
makes use of the difference in shape between the top and the base of
the part to be fed. Because of the width of the track and the wiper
blade, the cup-shaped part can only arrive at the cutout resting on
its base or on its top. It can be seen from the figure that the cutout
has been designed so that a part resting on its top falls off the track
and into the bowl, whereas one resting on its base passes over the
cutout and moves on to the delivery chute.

Figure 5.6 shows another application of a cutout where the area
covered by the top of a part is very much smaller than the area covered
by its base. In this case a V-shaped cutout rejects any part resting
on its top.

In Fig. 5.7 an example is shown where U-shaped parts are oriented.
With parts of this type, it is convenient to feed them supported on a

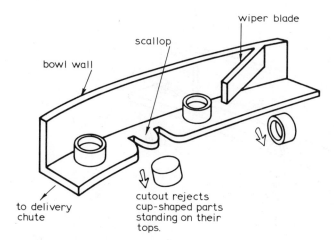

Fig. 5.5 Orientation of cup-shaped parts in vibratory bowl feeder.

Fig. 5.6 Orientation of truncated cones in vibratory bowl feeder.

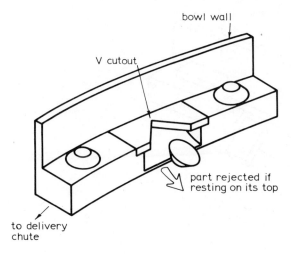

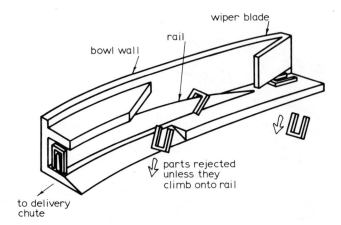

Fig. 5.7 Orientation of U-shaped parts in vibratory bowl feeder.

Fig. 5.8 Narrowed track.

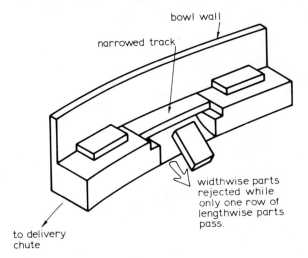

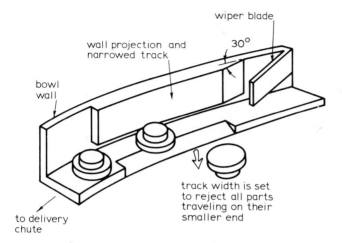

Fig. 5.9 Wall projection and narrowed track.

rail. In this case a proportion of the parts climbs onto the rail and passes to the delivery chute. The remainder either falls directly into the bowl or falls into the bowl through a slot between the rail and the bowl wall.

Figure 5.8 shows a narrowed track orienting device which is generally employed to orient parts lengthwise end to end, while permitting only one row to pass. Finally, shown in Fig. 5.9, is a wall projection and narrowed track device used to feed and orient parts with steps or grooves, such as short, headed parts. A short, headed part traveling on its larger end passes through the device, but other orientations are rejected back into the bowl.

5.2 Effect of Active Orienting Devices on Feed Rate

Sometimes, a part used on an assembly machine has only a single orientation, but more often the number of possible orientations is considerably greater. If, for example, a part had eight possible orientations and the probabilities of the various orientations were equal, and further, if only passive orienting devices were used to orient the parts, the feed rate of oriented parts would be only one-eighth of the actual feed rate. It is clear that if active orienting devices could be utilized, the feed rate of oriented parts could be considerably increased.

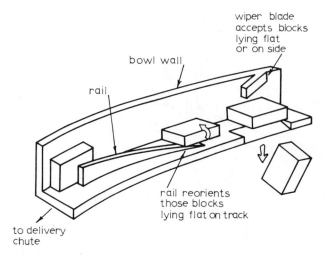

bowl wall

rail

wiper blade
accepts blocks
lying flat
or on side

rail reorients
those blocks
lying flat on track

to delivery
chute

Fig. 5.10 Orientation of rectangular blocks in vibratory bowl feeder.

To illustrate this point, consider the orienting system shown in Fig. 5.10 for feeding rectangular blocks. At one point the width of the track is such that blocks can only be fed with their long axes parallel to the direction of motion. Also, a wiper blade is arranged so that blocks lying flat or standing on their sides will be accepted. It is assumed in this example that the width of a block is less than twice its thickness. Finally, an active orienting device in the form of a tapered element ensures that blocks lying flat will be turned to stand on their sides. With this arrangement all the blocks fed up the track with their long axes parallel to the conveying direction will be fed from the bowl. If, however, the active orienting device was not part of the system and the wiper blade had been arranged to accept only the blocks lying flat on the track, a larger proportion of blocks would have been rejected, with a consequent reduction in feed rate.

5.3 Analysis of Orienting Systems

To determine the effect of certain aspects of part geometry on the efficiency with which it can be fed and oriented, several parts having the same basic shape but of different sizes can be considered. Figure 5.11 shows a family of parts that take the form of plain cylinders with a blind hole drilled axially from one end. These parts have the same diameter of 12.7 mm with an 11.7 mm diameter square bottom hole drilled from one end to a depth of 0.718 times the length of the parts.

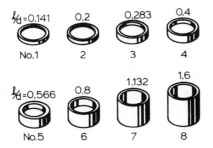

Fig. 5.11 Parts used in the experiment.

It can be shown that for all these parts, the center of mass is positioned at the bottom of the hole; the only geometric variable necessary to describe the part is its length-to-diameter ratio.

For a bowl feeder having a track that ensures single-file feeding of these parts, only four orientations of the part need to be considered. These orientations are shown in Fig. 5.12 and are lettered a, b_1, b_2, and c. Orientation a is when the part is fed standing on its base (that is, heavy end down). Orientations b_1 and b_2 are when the part is fed on its side, either heavy end first (b_1) or light end first (b_2). Finally, orientation c is when the part is fed heavy end up.

Before a study of the design of an orientation system for these parts can be made, it is necessary to know the probabilities with which these four orientations would initially occur. This information can then be used as the input to the orienting system analysis. Figure 5.13 presents the results of experimental and theoretical work. In the experiments, each part was repeatedly thrown onto a flat horizontal aluminum surface and the resulting final resting aspect was noted for

Fig. 5.12 Orientations of cup-shaped parts.

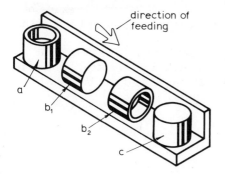

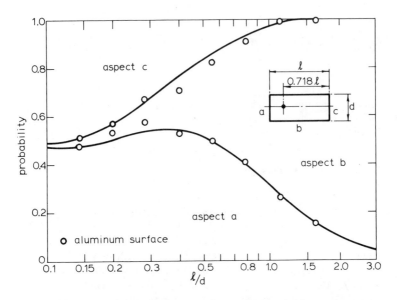

Fig. 5.13 Distribution of natural resting aspects.

each trial. In this experiment, it is not possible to distinguish between orientations b_1 and b_2 because the direction of feeding is not defined. For this reason, the term *natural resting aspect* is employed, which is meant to describe the way in which a part can rest on a horizontal surface. Thus, natural resting aspect a is when the part rests on its base, b is when the part rests on its side, and c is when the part rests with its heavy end up. Those parts that come to rest on their side will, when fed, divide about equally into the two orientations b_1 and b_2 described above. It can be seen from Fig. 5.13 that, for example, the probability that a cup-shaped part having a length-to-diameter ratio of 0.8 will come to rest on its base (natural resting aspect a) is 0.4. Each experimental point in Fig. 5.13 represents the results of at least 250 trials and is subject to 95% confidence limits of less than ±0.05.

The basis of a theoretical study of the distributions of natural resting aspects of parts is described later in this chapter. The solid lines shown in Fig. 5.13 are theoretical curves and can be seen to agree closely with the experimental results.

5.3.1 The Orienting System

A system for orienting the cup-shaped parts is shown in Fig. 5.14. It consists of one active device, a step; and two passive devices, a scallop and a sloped track with a ledge. The system is designed to deliver a part in orientation a (on its base).

In the orientation process, the parts first encounter the step device, whose object is to increase the proportion of parts in orientation a. This increase is achieved by arranging a step height that does not affect many of the parts in orientation a but reorients some of the parts in orientations b_1 or c into orientation a, as they pass over the step.

The remaining passive devices are simply designed to ensure that all parts remaining in orientations b_1, b_2, and c are rejected back into the bowl. These rejected parts later make a further attempt to filter through the orienting system.

The first of the passive devices, the scallop cutout, ensures rejection of a part in orientation c. Its design was based on data regarding the feeding motion of the part when the horizontal amplitude of vibration at the bowl wall was set at 1.0 mm with a vibration angle of 5 degrees. The part motion under these conditions is shown in Fig. 5.15 and was obtained using a computer program similar to that described by Redford and Boothroyd.* It can be seen from the figure that the part slides

*A. H. Redford and G. Boothroyd, "Vibratory Feeding," *Proc. I Mech. Eng.*, vol. 182, part 1, no. 6, 1967-1968.

Fig. 5.14 Experimental orienting system.

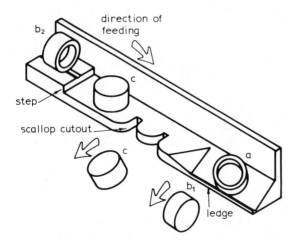

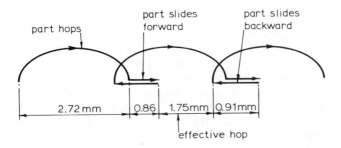

Fig. 5.15 Motion of part relative to track during experiments.

forward a distance of 0.86 mm after hopping and then slides backward
a distance of 0.91 mm before hopping again. Although the distance
the part hops is 2.72 mm, the maximum gap in the track that any point
on the undersurface of the part can negotiate is 1.75 mm. With this
information, it is possible to design the scallop device so that parts
in orientation a are always supported, regardless of where they are
situated on the device, whereas those parts in orientation c will, at
some point, be situated in a position where they cannot be supported
and will fall off the track into the bowl. Experiments showed that a
small proportion of those parts in orientations b_1 and b_2 are also rejected
by this device, but this does not affect the performance of the system
since the next device will reject parts in both of these orientations.

The second passive device, the sloped track with a ledge, is designed
to reject all those parts in orientations b_1 and b_2. The ledge retains
parts in orientation a but does not prevent parts in orientations b_1
and b_2 from rolling off the track and back into the bowl.

The only orienting system variable considered here is the height
of the active step orienting device, and in order to perform an optimal
design analysis, it is necessary to carry out an experimental program
where the effect of various step heights on each orientation, of each
of the eight specimens, is measured. Typical results of such experiments are presented in Fig. 5.16, which shows the effects of feeding
part 7 in each of its four initial orientations, over the step, with step
heights varying from zero to a maximum of 7 mm. It was found that
for step heights greater than 7 mm, the parts would bounce erratically
upon landing on the track below the step, an effect that would be unacceptable in practice.

5.3.2 Method of System Analysis

The object of an analysis of a bowl feeder orienting system is to
design each device so that the highest value for the efficiency of the
complete system is obtained. The efficiency of the system is defined

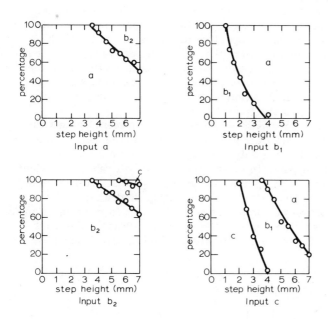

Fig. 5.16 Effect of step on part 7.

Fig. 5.17 Orienting system analysis.

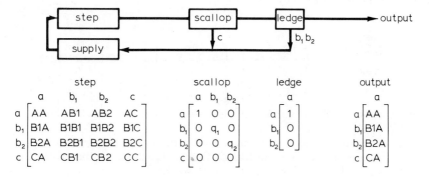

as the number of properly oriented parts delivered by the system, divided by the number of parts entering the system.

To calculate this efficiency for a system of orienting devices, a matrix technique has been developed.* Each device is represented by a matrix whose number of rows and columns depends on the number of orientations in the device's respective input and output. If these matrices are multiplied in the order that the parts encounter the devices, the resulting single column matrix represents the performance of the system. If this matrix is then premultiplied by a single row matrix representing the initial distribution of orientations of the part, the efficiency of the system is obtained.

Figure 5.17 shows a schematic diagram for the present system together with the appropriate matrices. The terms in the matrix that represent the step device are only symbolic. The term AA indicates the proportion of those parts in orientation a that will remain in orientation a, AB1 represents those parts in orientation a that are reoriented into b_1, and so on. In the matrix for the scallop device, q_1 and q_2 represent the proportion of parts that enter the device in orientation b_1 and b_2, respectively, and exit in the same orientation. For part 7 and a step height of 7 mm the step orienting device matrix becomes

$$
\begin{array}{c}
\quad\quad a \quad\ b_1 \quad\ b_2 \quad\ c \\
\begin{array}{c} a \\ b_1 \\ b_2 \\ c \end{array}
\begin{bmatrix}
0.50 & 0 & 0.50 & 0 \\
1.00 & 0 & 0 & 0 \\
0.30 & 0 & 0.64 & 0.06 \\
0.80 & 0.20 & 0 & 0
\end{bmatrix}
\end{array}
\tag{5.1}
$$

and the resulting system matrix is

$$
\begin{array}{c}
\quad a \quad\ b_1 \quad\ b_2 \quad\ c \\
\begin{array}{c} a \\ b_1 \\ b_2 \\ c \end{array}
\begin{bmatrix}
0.50 & 0 & 0.50 & 0 \\
1.00 & 0 & 0 & 0 \\
0.30 & 0 & 0.64 & 0.06 \\
0.80 & 0.20 & 0 & 0
\end{bmatrix}
\begin{bmatrix}
1 & 0 & 0 \\
0 & q_1 & 0 \\
0 & 0 & q_2 \\
0 & 0 & 0
\end{bmatrix}
\begin{bmatrix}
1 \\ 0 \\ 0 \\ 0
\end{bmatrix}
=
\begin{bmatrix}
0.50 \\ 1.00 \\ 0.30 \\ 0.80
\end{bmatrix}
\end{array}
\tag{5.2}
$$

This result means that 50% of those parts that enter the system in orientation a exit it in orientation a, 100% of those parts that enter in orientation b_1 exit in orientation a, and so on.

---

*L. E. Murch and G. Boothroyd, "Predicting Efficiency of Parts Orienting Systems," *Automation*, vol. 18, no. 2, Feb. 1971.

From Fig. 5.13 the initial distribution matrix for part 7 is

$$[a \quad b_1 \quad b_2 \quad c] = [0.27 \quad 0.35 \quad 0.35 \quad 0.03] \tag{5.3}$$

Thus, premultiplying the system matrix, Eq. (5.2), by the input distribution matrix, Eq. (5.3), gives

$$[0.27 \quad 0.35 \quad 0.35 \quad 0.03] \begin{bmatrix} 0.50 \\ 1.00 \\ 0.30 \\ 0.80 \end{bmatrix} = 0.61$$

which means that under these conditions, the efficiency of the system is 61%. Hence, if the bowl was set to feed parts at a rate of 10 parts per minute, the mean delivery rate of parts in orientation a would be 6.1 parts per minute.

5.3.3 Optimization

To optimize the design of this system, it is necessary to determine the step height that gives the maximum efficiency for each of the eight parts. The simplest method is to calculate the system efficiency for

Fig. 5.18 Effect of step height on efficiency (parts 1, 2, 3, and 4).

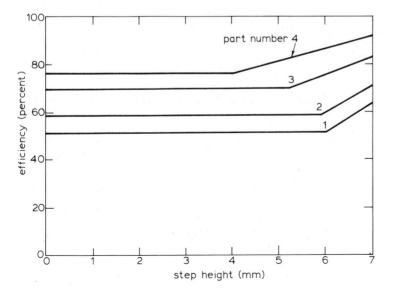

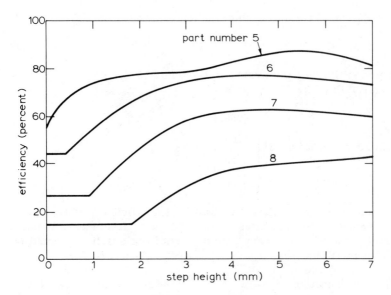

Fig. 5.19 Effect of step height on efficiency (parts 5, 6, 7, and 8).

Fig. 5.20 Effect of part shape on efficiency of orienting system.

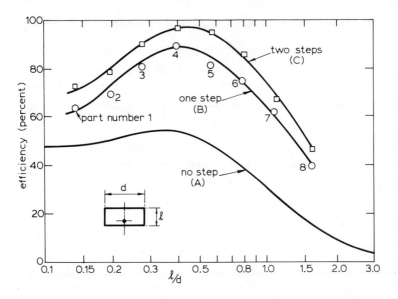

increments of step height within the practical range. The results of
this procedure are shown in Figs. 5.18 and 5.19, where it can be seen
that the maximum efficiency for five of the parts occurs at the maximum
allowable step height of 7 mm. These maxima are plotted in Fig. 5.20
(curve B) and compared with the initial distribution of parts in orientation a (curve A). Since this initial distribution is the same as the
system efficiency which would be obtained if the step device had not
been included in the system, the figures shows clearly the advantages
of including the step device. In many cases the efficiency of the system
is almost doubled; in other words, the output rate of oriented parts
would be almost doubled. Since such significant advantages of including
the step device in the system can be demonstrated, it is of interest to
consider the effect of including a further step device. In this case a
further variable is introduced and the upper curve shown in Fig. 5.20
is obtained. In all cases it is seen that the overall maximum efficiency
occurs for a part having a length-to-diameter ratio L/D of 0.4.

5.4 Performance of an Orienting Device

In the preceding section, to optimize the design of the orienting system
for cup-shaped parts it was necessary to determine the step height
giving the maximum efficiency. The information employed to determine
the performance of the step orienting device was empirical. It would
be most useful in investigations involving the optimization of an orienting system if theoretical expressions were available which described
with sufficient accuracy the performance of the orienting device. The
present section describes the analysis of one of the passive orienting
devices introduced earlier: a V cutout (Fig. 5.6) applied to the
orientation of truncated cone-shaped parts. It can be seen from Fig.
5.6 that, with a properly designed device, those parts being fed base
uppermost will be rejected, whereas those parts being fed on their
bases will be accepted and allowed to proceed to the outlet chute.
    Since the height of the part has no effect on the performance of
the device, the only parameters necessary to describe its important
characteristics are the radius of the base R and the radius of the top r.
The symmetrical orienting device may also be described completely by
only two parameters: the half angle of the cutout θ and the distance b
from the apex of the cutout to the bowl wall.
    During vibratory feeding the part proceeds along the track by a
combination of discrete sliding motions either backward or forward or
both and, under certain conditions, by a forward hop. All the motions
occur sequentially during each cycle of the vibratory motion of the
bowl. During each cycle, when the conditions are such that the part
hops, there will usually be a distance along the track. denoted by J,

where the part does not touch the track. Therefore, J is the smallest gap or slot in the track that will reject all particles that travel with this particular motion. For vibratory conditions that produce relatively small sliding motions when compared to the hop, the motion can be characterized by a series of equal hops, each a distance J.

The object of the design of a V cutout would be to determine the values of the parameters θ and b such that for a given part (given values of R and r) and for a given feeding characteristic (given J), all the parts fed on their tops would be rejected and a maximum of those fed on their bases would be accepted.

5.4.1 Analysis

Figure 5.21 shows two limiting conditions for the position of a part resting on its top. The first position is where the center of the part lies at P on the edge of the cutout. Thus, if the part comes to rest momentarily just to the right of P, it will be rejected. The second limiting condition is where the center of the part lies at Q and the edge of the part is just supported at D by the edge of the cutout. Thus, if the part contacts the track with its center anywhere between P and Q, it will be rejected. Two similar limiting conditions not shown in the figure will exist to the right of the cutout centerline and these positions may be deduced from the symmetry of the situation.

It is clear that for a part traveling from left to right (in the figure) in a series of hops, the probability that its center will fall in the space between P and Q (and thus be rejected) is given by j/J, where j is the distance between P and Q and J is the length of each hop.

Fig. 5.21 Determination of j for small cutout angles.

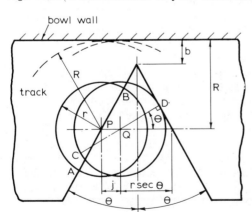

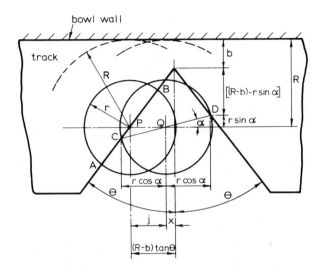

Fig. 5.22 Determination of j for large cutout angles.

For those parts that negotiate the gap between P and Q the probability that these parts will be rejected is the product of the probability that they will clear the first gap $(1 - j/J)$ and the probability that they will be rejected in the similar gap lying to the right of the cutout centerline (j/J). Hence, the total probability R_e that a part will be rejected in one of the two gaps is

$$R_e = \frac{j}{J} + (1 - \frac{j}{J}) \frac{j}{J} = 2(\frac{j}{J}) - (\frac{j}{J})^2 \qquad (5.4)$$

For the conditions illustrated in Fig. 5.21,

$$j = 2(R - b) \tan \theta - r \sec \theta \qquad (5.5)$$

For large cutout angles Eq. (5.5) does not always apply because when the top of the part is supported at D by the right-hand edge of the cutout, the point C, diametrically opposite to D, may be to the right of the left-hand edge of the cutout. Thus, the part in this situation will be rejected and an alternative limiting condition must be found. This is shown in Fig. 5.22. In this case, the point P is unchanged but the point Q is found by arranging that the diameter CD of the part is just supported between the edges of the cutout. From Fig. 5.22,

$$j = (R - b) \tan \theta - x \qquad (5.6)$$

and

$$x = r \cos \alpha - [(R - b) - r \sin \alpha] \tan \theta \qquad (5.7)$$

Also,

$$x = [(R - b) + r \sin \alpha] \tan \theta - r \cos \alpha \qquad (5.8)$$

Eliminating α from Eqs. (5.7) and (5.8) gives

$$x = \tan \theta [r^2 - (R - b)^2 \tan^2 \theta]^{1/2} \qquad (5.9)$$

Finally, substitution of Eq. (5.9) into Eq. (5.6) gives

$$j = \tan \theta \{(R - b) - [r^2 - (R - b)^2 \tan^2 \theta]^{1/2}\} \qquad (5.10)$$

The value of b at which Eq. (5.5) becomes invalid and Eq. (5.10) must be applied can be found by arranging that CD in Fig. 5.22 lies at right angles to the right-hand edge of the cutout. Under these conditions α becomes equal to θ and eliminating x from Eqs. (5.7) and (5.8) gives

$$(R - b) = r \cos \theta \cot \theta \qquad (5.11)$$

and thus from Eq. (5.5) or (5.10),

$$j = r(2 \cos \theta - \sec \theta) \qquad (5.12)$$

It can readily be shown that for $\theta \geq 45$ degrees, Eq. (5.10) always applies.

It is convenient to eliminate one variable from the foregoing expressions by dividing through by R and thereby writing the parameters in dimensionless form. Defining $r_0 = r/R$, $b_0 = b/R$, $j_0 = j/R$, and $J_0 = J/R$, the following equation is obtained for the rejection R_e of parts:

$$R_e = 2(\frac{j_0}{J_0}) - (\frac{j_0}{J_0})^2 \qquad (5.13)$$

where

$$j_0 = \tan \theta \{(1 - b_0) - [r_0^2 - (1 - b_0)^2 \tan^2 \theta]^{1/2}\} \qquad (5.14)$$

unless $\theta < 45$ degrees and $b_0 > (1 - r_0 \cos \theta \cot \theta)$, in which case

$$j_0 = 2(1 - b_0) \tan \theta - r_0 \sec \theta \qquad (5.15)$$

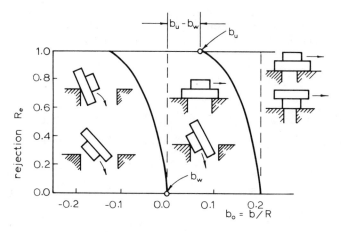

Fig. 5.23 Effect of b_0 on rejection of parts; $\theta = 30$ degrees, $J_0 = 0.15$.

These equations are presented in graphical form in Fig. 5.23 for a cutout having a half-angle of 30 degrees and a part where r_0 is 0.8.

The figure shows how the theoretical rejection rate R_e varies as the parameter b_0 is changed for a given value of the distance hopped J_0 by the part during each vibration cycle. The figures for the case when the part is being fed on its base were obtained by setting r_0 equal to unity. A negative b refers to the case where the apex of the cutout lies outside the interior surface of the bowl wall.

It can be seen from the figure that as the parameter b_0 is gradually decreased, a condition is eventually reached where the unwanted parts (those being fed on their tops) will start to be rejected. On further decreases in b_0, a point will be reached when all these parts will be rejected. Similar situations will arise for the wanted parts, but for lower values of b_0. Eventually, a value of b_0 will be reached where all the parts will be rejected. Clearly, in practice it will be necessary to choose a situation where all the unwanted parts will be rejected, even at the expense of rejecting some of the wanted parts.

After defining the largest value of b_0 for which all the unwanted parts are rejected as b_u and the smallest value of b_0 at which all the wanted parts are accepted as b_w, it can be stated that the best conditions would be those which resulted in the largest value of $(b_u - b_w)$. This would give the greatest working range for a given part and for given feeding conditions.

It is of interest to study how the magnitudes of b_u and b_w are affected by changes in the design parameters.

The value of b_u is obtained by setting R_e equal to unity in Eq. (5.13) with the appropriate equation for j_0, (5.14) or (5.15); the value of b_w is obtained by setting R_e equal to zero and r_0 equal to unity.

Thus, after rearrangement:

When $\theta \geq 45$ degrees:

$$b_u = 1 - \cos^2 \theta [J_0 \cot \theta + (r_0{}^2 \sec^2 \theta - J_0{}^2)^{1/2}] \qquad (5.16)$$

$$b_w = 1 - \cos \beta \qquad (5.17)$$

When $\theta \leq 45$ degrees:

$$b_u = 1 - 0.5 J_0 \cot \theta - 0.5 r_0 \operatorname{cosec} \theta \qquad (5.18)$$

$$b_w = 1 - 0.5 \operatorname{cosec} \theta \qquad (5.19)$$

unless $b_u < (1 - r_0 \cos \theta \cot \theta)$, in which case it is given by Eq. (5.16) and b_w is given by Eq. (5.19). The equations above are plotted in Fig. 5.24 and illustrate the effects of θ and J_0.

This theory has been developed for an idealized situation where the part proceeds along the track by hopping. However, in reality, both hopping and sliding occurs. It can be shown that although this would affect Eq. (5.13), and hence the shapes of the curves in Fig. 5.23, the values of b_u and b_w would be unchanged.

Values of the working range ($b_u - b_w$) obtained from Eqs. (5.16) through (5.19) are plotted against θ in Fig. 5.25. It can be seen that for larger values of J_0, an optimum condition exists giving the maximum working range. Further, at low values of θ, the magnitude of the working range becomes very sensitive to changes in J_0. Since most vibratory feeders operate at the same frequency, a large value of J_0 implies a high conveying velocity. In practice, it would clearly be desirable to choose conditions that give minimum sensitivity to changes in the feeding parameters and yet give the maximum working range for a reasonably large value of J_0.

The procedure used to obtain the experimental points for b_u and b_w, which are presented in Fig. 5.24, is outlined in Boothroyd and Murch.* It is seen in Fig. 5.24 that the results for b_u show good agreement with the theory over the whole range of cutout angles when J_0 is set equal to 0.15. The experimental values for b_w show good agreement with the theory only at small cutout angles. For larger cutout angles, the experimental value is always the larger.

---

*G. Boothroyd and L. E. Murch, "Performance of an Orienting Device Employed in Vibratory Bowl Feeders," Trans. ASME, *J. Eng. Ind.*, Aug. 1970.

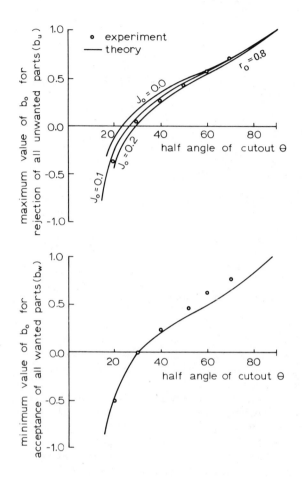

Fig. 5.24 Effect of θ on values of b_u and b_w.

Ideally, in the design of a V cutout orienting device, the pertinent data regarding the vibrating motion of the bowl feeder could be used to estimate the value of J_0 employing the results of Chapter 3. Subsequently, using Fig. 5.25, the half angle of the cutout θ could be chosen which gives the best value for the working range. From this figure, when $J_0 \leq 0.1$, the smaller the value of θ, the larger the working range. However, small cutout angles can present a practical problem. Under these circumstances, parts that are rejected may not be deflected properly from the bowl track and may interfere with the behavior of the following parts. Thus, the angle chosen should be that which gives the best working range and yet provides for adequate deflection of rejected parts.

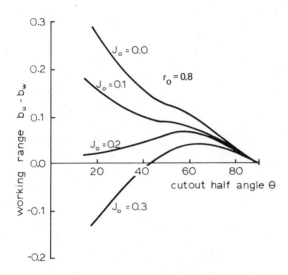

Fig. 5.25 Effect of cutout angle on the working range.

Since it is essential that all the unwanted parts be rejected, the value of b_0 (which defines the position of the cutout apex) must be less than b_u. The value of b_u can be found from either Eq. (5.16) or (5.18), whichever is appropriate.

In practice, this method will result in an effective orienting device, but not necessarily the most efficient one. If for the larger cutout angles, the experimental values of b_w were significantly larger than the theory predicted, the corresponding working ranges ($b_u - b_w$) would be negative. Thus, when the cutout is designed so that all the unwanted parts are rejected, some of the wanted parts will also be rejected. This reduces the output and can result in a low efficiency.

In view of these observations, the recommended procedure would be to choose a value of θ less than 45 degrees but large enough to give acceptable deflection and then determine b_u from the appropriate equation. Such an orienting device would have a positive working range when J_0 is less than 0.2 and thus an efficiency of 100%.

5.4.2 Use of the Results

Based on the work described here, a design data sheet for the V cutout is shown in Fig. 5.26, which is taken from Boothroyd et al.*

---

*G. Boothroyd, C. Poli, and L. E. Murch, "The Handbook of Feeding and Orienting Techniques for Small Parts," Department of Mechanical Engineering, University of Massachusetts, Amherst, Mass., 1977.

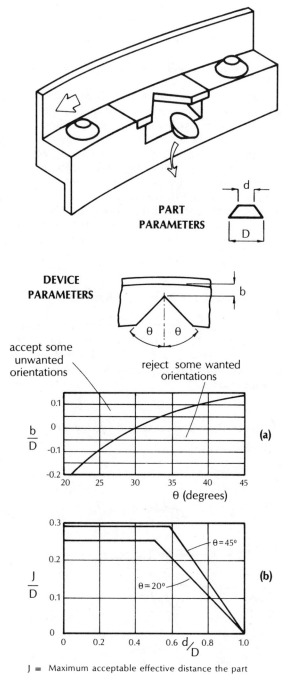

PART PARAMETERS

DEVICE PARAMETERS

accept some unwanted orientations

reject some wanted orientations

$\frac{b}{D}$

θ (degrees)

(a)

$\frac{J}{D}$

θ = 45°

θ = 20°

d/D

(b)

J = Maximum acceptable effective distance the part hops during each vibration cycle.

The V cut-out orienting device is a triangular notch in the track that is used to reject a part traveling on its small circular top. A part traveling on its larger circular base will pass over the device.

The important design parameters of the device are the half-angle of the cut-out θ and the distance b from the apex of the triangle to the bowl wall. The magnitudes of these parameters can be determined from Fig. (a).

The device is completely effective if the effective distance the part hops J is less than the value determined using Fig. (b). This effective hop is a function of the geometry of the feeder as well as its amplitude and frequency of vibration. Values of J for given feeding conditions can be obtained from section 4 of this manual.

Example:

Given D = 40 mm and d = 32 mm, suppose θ = 35°. Then from Fig. (a)

$$\frac{b}{D} = 0.07$$

and

$$b = 2.8 \text{ mm}$$

Also since $\frac{d}{D} = 0.8$ then from Fig. (b)

$$\frac{J}{D} = 0.13$$

and therefore

$$J = 5.2 \text{ mm}$$

Fig. 5.26 V cutout data sheet. (Adapted from G. Boothroyd, C. Poli, and L. E. Murch, "The Handbook of Feeding and Orienting Techniques for Small Parts," Department of Mechanical Engineering, University of Massachusetts, Amherst, Mass., 1977.)

An example of the design of this device using this information is also shown in Fig. 5.26.

Figure 5.26a is based on Eq. (5.16) or (5.18). Figure 5.26b is based on the assumptions that, first, the working range ($b_u - b_w$) should be positive so that wanted parts will not be rejected, and second, the hop J must be considerably less than 0.5D so that the cutout will function and so that the probability of a part being rejected on the second gap is independent of its probability of rejection at the first gap. Thus, Fig. 5.26b is obtained by setting b_u equal to b_w and conservatively limiting the value of J/D as shown.

"The Handbook of Feeding and Orienting Techniques for Small Parts" also contains design data sheets for the other orienting devices discussed in this chapter and should be consulted when designing orienting systems.

5.5 Natural Resting Aspects of Parts for Automatic Handling

In order to analyze the complete system for orienting cup-shaped parts (Fig. 5.14) it was necessary to know the probabilities with which the various orientations would initially occur, and Fig. 5.13 presented the experimental and theoretical results of the distribution of the natural resting aspects for cup-shaped parts.

Fig. 5.27 Simple orienting system for a vibratory bowl feeder ($L/D = 0.7$).

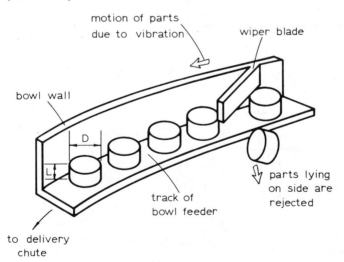

motion of parts due to vibration

wiper blade

bowl wall

D

L

track of bowl feeder

parts lying on side are rejected

to delivery chute

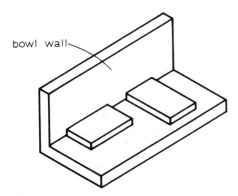

Fig. 5.28 Rectangular parts on a track.

Figure 5.27 shows another orienting system for a vibratory bowl feeder. The parts being fed and oriented are cylinders whose length-to-diameter ratio is 0.7. The wiper blade is adjusted to reject parts lying on their sides and to allow parts lying on end to pass to the delivery chute. Clearly, the feed rate of oriented parts will depend on the rate at which parts encounter the wiper blade and the proportion of these parts that are lying on end. For the case shown in Fig. 5.27, the proportion of parts lying on end is, surprisingly, only about 0.3. This means that only 30% of parts fed to the wiper blade will pass through to the delivery chute.

From the preceding example it is seen once again that knowledge of the probabilities of the various ways the part will naturally rest (natural resting aspects) is essential in any analysis of the performance of the orienting system. In some cases it is also necessary to know how a particular natural resting aspect will divide into various orientations on the bowl track. The difference between natural resting aspect and orientation is illustrated in Fig. 5.28. Here, both parts have the same natural resting aspect (that is, lying on their sides) but have different orientations on the bowl track.

5.5.1 Assumptions

Repeated observations and analyses of the behavior of a variety of parts dropped onto different surfaces have led to the following conclusions.

1. Surfaces can be divided into two main categories. First, there is the soft rubberlike type of surface (referred to as soft surfaces) where, on impact, a corner of the part digs into the surface, result-

ing in an impact force having a significant horizontal component in addition to a vertical component. The actions of these force components and the nature of contact between the part and the surface cause the part to roll across the surface, changing rapidly from one natural resting aspect to another.

The secondary category of surface is hard and resilient and includes such materials as metals, glass, and hard laminated plastics. With surfaces of this nature (referred to as hard surfaces) the corners of the part do not generally dig into the surface and the horizontal component of the impact force is negligible in its effect. After impact, the part does not usually roll across the surface, but bounces up and down, overturning repeatedly, but remaining generally in the same area on the surface. Thus, a change of aspect must be brought about by vertical impact forces applied at the edges and corners of the part.

2. The probability that a part will come to rest in a particular natural resting aspect is a function of two factors: (1) the energy barrier tending to prevent a change of aspect; and (2) the amount of energy possessed by the part when it begins to fall into that natural resting aspect.

3. It will be assumed throughout that parts are dropped from a height sufficient to ensure that after impact, at least one change in natural resting aspect occurs.

5.5.2 Analysis for Soft Surfaces

Consider a square prismatic part to be initially resting on a flat horizontal surface with one of its corners, say A, at the origin of a rectangular XY coordinate system, which also lies on the surface. Figure 5.29 shows this part rotated about the X axis through some angle θ small enough so that if it possessed no kinetic energy, it would fall onto its end. To move onto its side by rolling over corner A, such that the part rotates about a line parallel to the Y axis, the part would require sufficient energy to raise its center of mass from its initial position B to position C when the part is just about to change its aspect.

From the projections in Fig. 5.29 it can be seen that this change of height is equal to

$$BC = [x^2 + \ell^2 \cos^2 (\beta - \theta)]^{1/2} - \ell \cos (\beta - \theta) \qquad (5.20)$$

where ℓ is given by $(x^2 + y^2)^{1/2}$ and β is given by arctan (x/y). This expression gives the length of the vertical line BC in Fig. 5.29b, which is the projection of points B and C in Fig. 5.29c onto a line perpendicular to the XY plane. If all such lines (as θ varies from $-\beta$ to $+\beta$) were drawn on the projection in Fig. 5.29b, the shaded area shown in Fig. 5.30a would result. This shaded area represents the total energy barrier for the part resting on end.

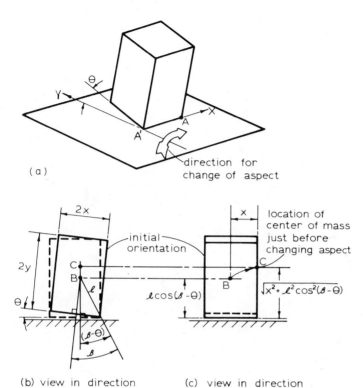

direction for
change of aspect

(b) view in direction
of arrow X

(c) view in direction
of arrow y

Fig. 5.29 Energy required to change aspect of square prism tilted
at an angle θ.

Fig. 5.30 Energy barriers for a square prism.

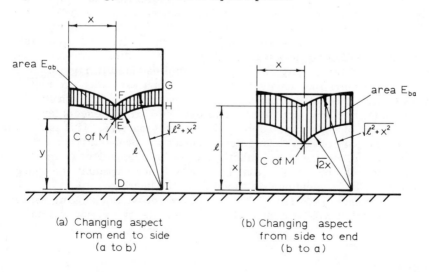

(a) Changing aspect
from end to side
(a to b)

(b) Changing aspect
from side to end
(b to a)

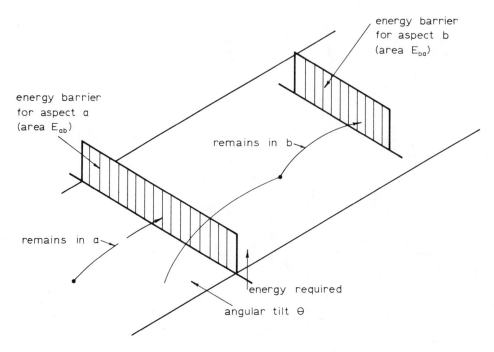

energy barrier
for aspect b
(area E_{ba})

energy barrier
for aspect a
(area E_{ab})

remains in b

remains in a

energy required

angular tilt θ

Fig. 5.31 Representation of energy barriers showing that the number of parts remaining in an aspect will be proportional to the energy barrier for that aspect.

The size of the area can be obtained by integrating Eq. (5.20) between the limits $\theta = \pm\beta$. Alternatively, the area can be obtained from Fig. 5.30 by geometry.

The present theory is based on the hypothesis that the probability that a part will come to rest in a particular orientation is proportional to the area of the energy barrier. This is illustrated in Fig. 5.31, which represents the energy barriers for a part with only two resting aspects (say a regular prism or cylinder that can rest only on its side or on end). As the part tumbles end over end, it will encounter, on each change of aspect, the appropriate energy barrier. If it is assumed that as the part passes through each aspect the probability that it possesses a given amount of kinetic energy is independent of the angle of tilt, then the probability that the part will not surmount the energy barrier will be proportional to the area of the barrier.

Thus, referring to Fig. 5.31, the number of parts changing aspect from a to b that will be stopped by the first energy barrier will be proportional to the area of the energy barrier E_{ab}. Similarly, the number stopped by the second energy barrier will be proportional to E_{ba}. This situation will continue until the part comes to rest.

Thus, in the end, the number of parts N_a and N_b remaining in aspects a and b will be given, respectively, by

$$N_a \propto E_{ab}$$
$$N_b \propto E_{ba}$$

(5.21)

Hence, returning to the example in Fig. 5.30, by geometry

$$E_{ab} = 2(\text{area enclosed by points EFGH})$$

or

$$E_{ab} = 2(A_{DFI} + A_{IFG} - A_{DEI} - A_{IEH})$$

(5.22)

where A_{DFI} refers to the area enclosed by points D, F, I; A_{IFG} refers to the area enclosed by points I, F, G; and so on.

Thus, it can be shown that

$$E_{ab} = x^2(\alpha_2 p^2 + q - \alpha_1 q^2 - \frac{y}{x})$$

(5.23)

and, similarly,

$$E_{ba} = x^2(\alpha_2 p^2 + q - \frac{\pi}{2} - 1)$$

(5.24)

where

$$p = \frac{\sqrt{\ell^2 + x^2}}{x} = \sqrt{2 + (\frac{y}{x})^2}$$

$$q = \frac{\ell}{x} = \sqrt{1 + (\frac{y}{x})^2}$$

$$\alpha_1 = \arcsin (\frac{1}{q})$$

$$\alpha_2 = \arcsin (\frac{1}{p})$$

Since the part has two ends and four sides, the probability for aspect a (on end) is

$$P_a = \frac{2E_{ab}}{2E_{ab} + 4E_{ba}} \tag{5.25}$$

and the probability for aspect b (on side) is

$$P_b = 1 - P_a \tag{5.26}$$

Calculation of the values of E_{ab} and E_{ba} from Eqs. (5.23) and (5.24) and substitution of Eqs. (5.25) and (5.26) give the curve shown in Fig. 5.32. Also shown in this figure are experimental results for steel parts dropped onto a rubber-coated (soft) surface.

The energy barrier for a solid cylindrical part is obtained in a similar manner. Figure 5.33a shows such a part that was initially resting on end on a flat horizontal surface but has been rotated about the X axis through a small angle θ so that it would fall onto its end if it were released. To move the cylinder onto its side by now rotating it about the Y axis causes the center of mass to rise further as shown by the energy barrier in Fig. 5.34a.

If the part were first lying on its side, as shown in Fig. 5.33b, and then rolled on the horizontal surface, the distance from the center

Fig. 5.32 Probabilities of natural resting aspects for square prisms dropped onto a soft surface.

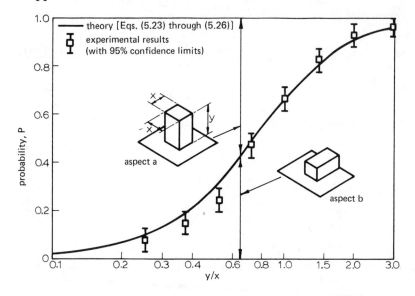

(a)

(b)

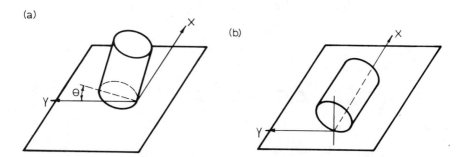

Fig. 5.33 Solid cylindrical part tilted through an angle θ (a) and solid cylindrical part lying on its side (b).

of mass to the surface would remain constant. If the part were now rotated onto one end about one of its edges, the energy barrier becomes that shown in Fig. 5.34b.

In these cases the expressions for the areas of the energy barriers are rather simpler than in the case of a square prism, and

$$E_{ab} = \ell^2[(2 - \cos \alpha) \sin \alpha - \alpha] \qquad (5.27)$$

$$E_{ba} = \ell^2(1 - \sin \alpha)\pi \sin \alpha \qquad (5.28)$$

Fig. 5.34 Energy barriers for a cylinder on a soft surface.

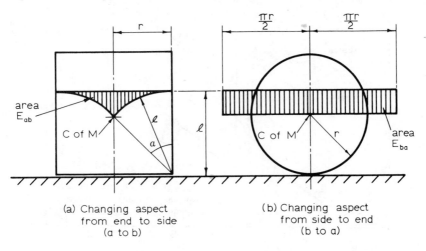

(a) Changing aspect
from end to side
(a to b)

(b) Changing aspect
from side to end
(b to a)

where $\alpha = \cot (L/D)$; L is the length and D is the diameter of the part. Hence,

$$P_a = \frac{E_{ab}}{E_{ab} + E_{ba}}$$

$$= \frac{(2 - \cos \alpha) - (\alpha/\sin \alpha)}{(2 - \cos \alpha) - (\alpha/\sin \alpha) + \pi(1 - \sin \alpha)} \qquad (5.29)$$

and

$$P_b = 1 - P_a \qquad (5.30)$$

Rather surprisingly, it is found that if the results for a square prism (Fig. 5.32) are plotted in terms of the L/D ratio of the circumscribed cylinder, they can be very closely approximated by the results in Fig. 5.35 for a solid cylinder. Indeed, application of the theory to prisms of any regular cross section (triangular, square, pentagonal, hexagonal, etc.), together with experiments, shows that Eqs. (5.27) through (5.30) apply to any solid regular prism, where L/D is defined as the length-to-diameter ratio of the circumscribed cylinder. This is illustrated in Fig. 5.36.

Fig. 5.35 Probabilities of natural resting aspects for cylinders dropped onto a soft surface.

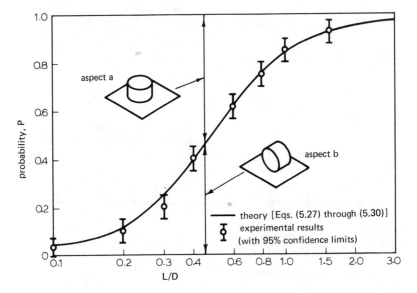

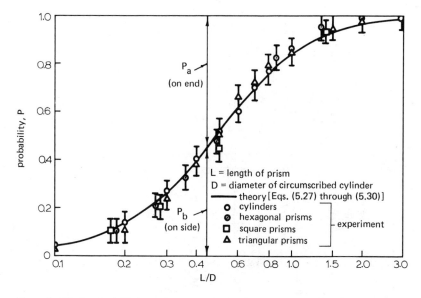

Fig. 5.36 Probabilities of natural resting aspects for cylinders and various regular prisms dropped onto a soft surface.

5.5.3 Analysis for Hard Surfaces

For hard surfaces, the approach described in Boothroyd et al.* can be used in conjunction with the new energy-barrier technique described above. In this case it is assumed that an extra amount of energy is required to change a part aspect, because energy can be provided only by a vertical impact force applied at one of the edges or corners of the part. In addition, allowance must be made for the effect of this additional energy, which tends to keep the part from remaining in the next aspect.

Again, as with the soft-surface solution and experiments, it was found that the results for a solid cylinder could be applied to any prism of regular cross section with three or more sides that could be enclosed within the cylinder.

---

*G. Boothroyd, A. H. Redford, C. Poli, and L. E. Murch, "Statistical Distribution of Natural Resting Aspects of Parts for Automatic Handling," *Manuf. Eng. Trans.*, vol. 1, 1972.

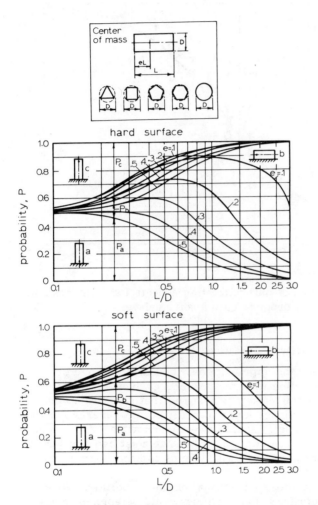

Fig. 5. 37 Probabilities of natural resting aspects for prisms of regular cross section. L is the length of prism and D the diameter of circumscribed cylinder. (Adapted from G. Boothroyd, C. Poli, and L. E. Murch, "The Handbook of Feeding and Orienting Techniques for Small Parts," Department of Mechanical Engineering, University of Massachusetts, Amherst, Mass., 1977.)

5.5.4 Analysis for Cylinders and Prisms with Displaced Centers of Mass

If the center of mass of a cylinder is displaced along its axis from the midpoint, the probabilities for the cylinder to come to rest on either of its ends become different. Based on experimental and theoretical work, it was found that the approach described above for both hard and soft surfaces could be applied equally successfully to this problem.

5.5.5 Summary of Results

Figure 5.37 presents a complete summary of the results of this work. It can be seen that the probabilities for the natural resting aspects of all prisms of regular cross section can be presented in only two graphs (one for a soft surface and one for a hard surface). This figure takes the form of a data sheet designed for inclusion in "The Handbook of Feeding and Orienting Techniques for Small Parts."

5.6 Analysis of a Typical Orienting System

In general, an orienting system for a vibratory bowl feeder consists of one or more orienting devices arranged in series along the bowl track. These devices are usually located near the outlet of the feeder

Fig. 5.38 Orienting system for right rectangular prisms.

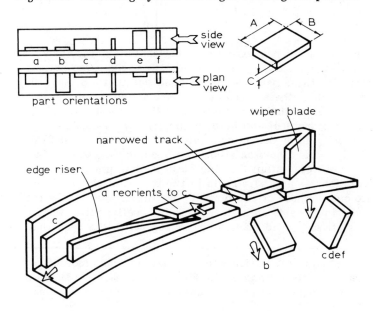

along a horizontal portion of the track. Because the track section is level, the parts can travel at a conveying velocity that is greater than the velocity of the parts on the preceding incline, thus enabling parts to separate and eliminating interference between adjacent parts as they pass the various devices. Such a system is shown in Fig. 5.38 for the feeding and orienting of right rectangular prisms. The six orientations for these prisms are also described in this figure.

The first device is a wiper blade, and it rejects orientations c, d, e, and f back into the bowl. It also serves to remove the secondary layers of parts where one part rests on another instead of on the track. The output of this device is either orientation a or b.

The narrow track is next, and it rejects orientation b back into the bowl, leaving only orientation a. The riser turns orientation a into orientation c, which is the output orientation of the system.

The matrices for these devices and systems are:

wiper blade　　narrow track　　riser　　system

$$
\begin{array}{c}
\text{wiper blade} \\
\begin{array}{c} a \quad b \end{array} \\
\begin{array}{c}a\\b\\c\\d\\e\\f\end{array}
\begin{bmatrix}1&0\\0&1\\0&0\\0&0\\0&0\\0&0\end{bmatrix}
\end{array}
\qquad
\begin{array}{c}
\begin{array}{c} a \end{array}\\
\begin{array}{c}a\\b\end{array}
\begin{bmatrix}1\\0\end{bmatrix}
\end{array}
\qquad
\begin{array}{c}
\begin{array}{c} c \end{array}\\
a[1] =
\end{array}
\qquad
\begin{array}{c}
\begin{array}{c} c \end{array}\\
\begin{array}{c}a\\b\\c\\d\\e\\f\end{array}
\begin{bmatrix}1\\0\\0\\0\\0\\0\end{bmatrix}
\end{array}
$$

Rectangular prisms, when tossed on a horizontal surface, can come to rest on one of three faces. These positions are the three natural resting aspects for these parts. The probabilities that a prism will come to rest in each of these three aspects are shown in Fig. 5.39. The parts used in this particular system are 45 × 30 × 3 mm. The values c/a and c/b are 0.07 and 0.10, respectively. According to Fig. 5.39, virtually all of these parts will rest on their largest face when tossed onto a hard horizontal surface such as the bottom of a bowl feeder. However, within this one natural resting aspect there are two orientations (a and b). Parts in the bottom of the bowl tend to rotate into one of these orientations before they travel up the inclined track. For aluminum parts on a steel track the coefficient of friction μ is 0.04. Thus, from Ho and Boothroyd* or Fig. 5.40 it is seen that

---

*C. Ho and G. Boothroyd, "Orientation of Parts on the Track of a Vibratory Feeder," Proceedings of the Fifth North American Metalworking Research Conference, Society of Manufacturing Engineers, Dearborn, Michigan, 1977, p. 363.

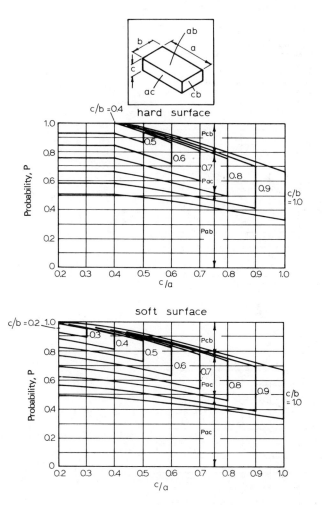

Fig. 5.39 Probabilities of natural resting aspects for right rectangular prisms. (Adapted from G. Boothroyd, C. Poli, and L. E. Murch, "The Handbook of Feeding and Orienting Techniques for Small Parts," Department of Mechanical Engineering, University of Massachusetts, Amherst, Mass., 1977.)

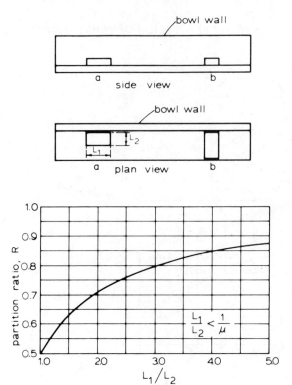

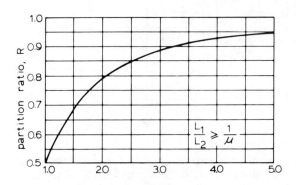

Fig. 5.40 Probabilities of orientations of regular prisms and rectangular prisms on a bowl feeder track. P_a is the probability of orientation a, P_b the probability of orientation b, partition ratio $R = P_a/(P_a + P_b)$, μ is the coefficient of friction between parts and bowl track, and L_1 and L_2 the length of long and short sides, respectively, of the part-track interface ($L_1 \geq L_2$). (From C. Ho and G. Boothroyd, "Orientation of Parts on the Track of a Vibratory Feeder," Proceedings of the Fifth North American Metalworking Research Conference, Society of Manufacturing Engineers, Dearborn, Michigan, 1977, p. 363.

for these parts the partition ratio $R = P_a/(P_a + P_b)$ is 0.63, and therefore the ratio of the probabilities is given by

$$P_a = R(P_a + P_b) = 0.63, \quad P_b = 0.37$$

The initial distribution matrix (IDM) showing the probable distribution
of the various orientations is therefore

a   b   c   d   e   f

[0.63 0.37 0 0 0 0]

Thus, the efficiency η of this system, which is given by the product
of the IDM and the system matrix, is 63%.

  The average length of a part entering this system is the product
of the IDM and a matrix of the lengths of the corresponding orientations in the conveying direction. For these parts the average part
length $\bar{l}$ is

$$[0.63 \quad 0.37 \quad 0 \quad 0 \quad 0 \quad 0] \begin{bmatrix} 45 \\ 30 \\ 45 \\ 3 \\ 30 \\ 3 \end{bmatrix} = 39 \text{ mm}$$

The feed rate F can be found from

$$F = \eta \frac{v}{\bar{l}} \tag{5.31}$$

where v is the conveying velocity of the parts on the inclined section
of the track when adjacent parts are touching. The feed rate of any
system can be found in a similar manner using the initial distribution
matrix, system matrix, length matrix, and Eq. (5.31). The feed rate
can also be determined from

$$F = \frac{vE}{A} \tag{5.32}$$

where E is known as the modified efficiency of the orienting system
and is given by

$$E = \frac{\eta A}{\bar{l}} \tag{5.33}$$

It is this modified efficiency that is presented on the data sheets in "The Handbook of Feeding and Orienting Techniques for Small Parts" and which simplifies their use.

5.6.1 Design of Orienting Devices

To achieve the calculated 63% efficiency, the orienting devices that make up the system must be properly designed. This is best done by using the design data and performance curves for orienting devices used in vibratory bowl feeders provided in Section 5 of "The Handbook of Feeding and Orienting Techniques for Small Parts."

Fig. 5.41 Wiper blade design. (Adapted from G. Boothroyd, C. Poli, and L. E. Murch, "The Handbook of Feeding and Orienting Techniques for Small Parts," Department of Mechanical Engineering, University of Massachusetts, Amherst, Mass., 1977.)

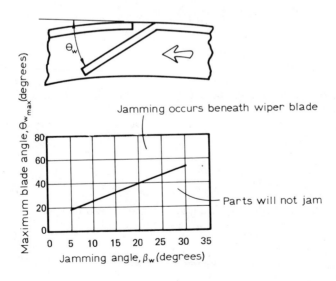

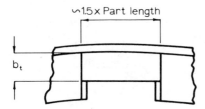

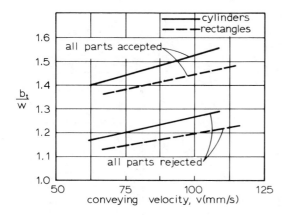

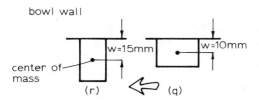

Fig. 5.42 Narrow-track design. (Adapted from G. Boothroyd,
C. Poli, and L. E. Murch, "The Handbook of Feeding and Orienting
Techniques for Small Parts," Department of Mechanical Engineering,
University of Massachusetts, Amherst, Mass., 1977.)

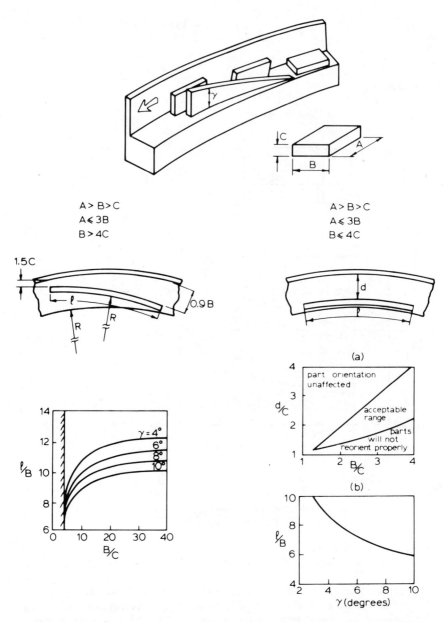

Fig. 5.43 Design data for the edge-riser orienting device. (Adapted from G. Boothroyd, C. Poli, and L. E. Murch, "The Handbook of Feeding and Orienting Techniques for Small Parts," Department of Mechanical Engineering, University of Massachusetts, Amherst, Mass., 1977.)

For the wiper blade, which rejects orientations c, d, e, and f
(Fig. 5.38), the angle between the wiper blade and the bowl wall θ
is set to avoid the jamming action produced by overlapping parts, as
shown in Fig. 5.41. The smallest jamming angle β_w for these parts
is arctan (3/45) or 3.8 degrees (0.07 rad), and the maximum value of
θ_w, from Fig. 5.41, is 18 degrees (0.31 rad). The height of the wiper
blade should be sufficient to remove a secondary layer of parts or 5 mm.

The narrow-track device rejects orientation b but allows orientation
a to pass. From Fig. 5.42 and a conveying velocity of 100 mm/s, the
corresponding values for the dimensionless track width b_t/w are 1.2
and 1.45, respectively. Thus, in millimeters

$$(1.2)(22.5) > b_t > (1.45)(15)$$

or

$$27 > b_t > 22$$

The narrow track should be 23 mm wide and 45 mm long.

The edge-riser orienting device turns orientation a into orientation
c. The design information for this device is presented in Fig. 5.43.
For a 6 degree (0.10 rad) riser angle and parts 45 × 30 × 3 mm, B/C
equals 10 and the ramp length is 300 mm. Other riser angles and
lengths will also produce satisfactory results.

5.7 Out-of-Bowl Tooling

A further type of orienting device is that which is situated between
the feeder and the workhead. Such devices are usually of the active
type because orientation by rejection is not often practicable. Figure
5.44 illustrates a device described by Tipping* where the position of
the center of gravity of a part is utilized. In this example the cup-
shaped part is pushed onto a bridge and the weight of the part acting
through the center of gravity pulls the part down nose first into the
delivery chute. In Fig. 5.45, the same part is reoriented using a
different principle. With this method, if the part passes nose first
down the delivery tube, it is deflected directly into the delivery chute
and maintains its original orientation. A part fed open-end first will
be reoriented by the pin located in the wall of the device.

The device illustrated in Fig. 5.46 is known as a selector and em-
ploys a principle that has been applied successfully to reorient a wide

---

*W. V. Tipping, "Mechanized Assembly Machines, 9: Orientation and
Selection," *Mach. Des. Eng.*, Feb. 1966, p. 38.

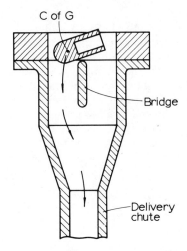

C of G

Bridge

Delivery
chute

Fig. 5.44 Reorientation of cup-shaped part. (C of G, center of gravity.) (From W. V. Tipping, "Mechanized Assembly Machines, 9: Orientation and Selection," *Mach. Des. Eng.*, Feb. 1966, p. 38.)

Fig. 5.45 Reorientation of cup-shaped part. (From W. V. Tipping, "Mechanized Assembly Machines, 9: Orientation and Selection," *Mach. Des. Eng.*, Feb. 1966, p. 38.)

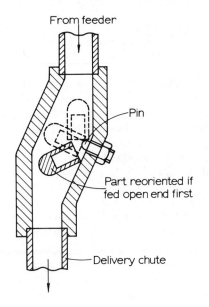

From feeder

Pin

Part reoriented if
fed open end first

Delivery chute

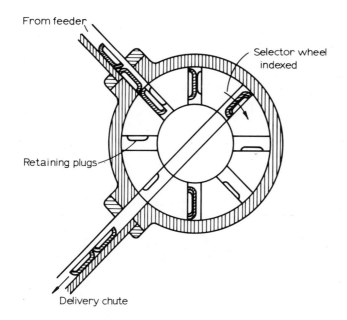

From feeder

Selector wheel
indexed

Retaining plugs

Delivery chute

Fig. 5.46 Reorientation of shallow-drawn parts. (From "Hopper Feeds as an Aid to Automation," Machinery's Yellowback No. 39, Machinery Publishing Co. Ltd., Brighton, England.)

variety of parts.* The selector consists of a stationary container in which is mounted a wheel with radial slots. The wheel is driven by an indexing mechanism to ensure that the slots always align with the chutes. In the design illustrated the shallow-drawn parts may enter the slot in the selector wheel in either of two attitudes. After two indexes the parts are aligned with the delivery chute and those which now lie open end upward slide out of the selector and into the delivery chute. Those which lie open end downward are retained by a plug in the slot. After a further four indexes the slot is again aligned with the delivery chute. The part has now been turned over and is free to slide into the delivery chute.

---

*"Hopper Feeds as an Aid to Automation," Machinery's Yellowback No. 39, Machinery Publishing Co. Ltd., Brighton, England.

Chapter 6

Feed Tracks, Escapements, Parts-Placing Mechanisms, and Robots

To provide easy access to automatic workheads and the assembly machine, the parts feeder is usually placed some distance away from the workhead. The parts, therefore, have to be transferred and maintained in an orientation between the feeder and the workhead by use of a feed track. Most parts feeders do not supply parts at the discrete intervals usually required by an automatic workhead. As a result, the parts feeder must be adjusted to overfeed slightly and a metering device, usually referred to as an escapement, is necessary to ensure that parts arrive at the automatic workhead at the correct intervals. After leaving the escapement, the parts are then placed in the assembly, a process usually carried out by a parts-placing mechanism.

6.1 Gravity Feed Track Arrangements

Feed tracks may be classified as either gravity tracks or powered tracks. The majority of tracks are of the gravity type and these may take many forms. Two typical track arrangements are illustrated in Fig. 6.1; the choice of design generally depends on the required direction of entry of the part into the workhead. In design, it should be remembered that the track may not always be full and it is desirable that feeding should still take place under this condition.

When the track is partly full and no pushing action is obtained by air jets or vibration, it is clear that the vertical delivery track design shown in Fig. 6.1b will deliver parts from rest at a greater rate than the horizontal delivery design shown in Fig. 6.1a. The performance of the vertical delivery track will also be independent of the loading

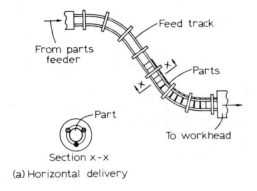

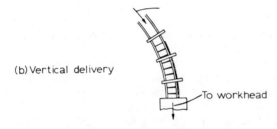

Fig. 6.1 Gravity feed track arrangements.

in the track, and if no further parts are fed into the track it will deliver the last part as quickly as the first. The time of delivery t_p will be given by the time taken for a part to fall a distance equal to its own length.

Thus

$$t_p = (\frac{2L}{g})^{1/2} \tag{6.1}$$

where L is the length of the part and g the acceleration due to gravity, or 9.81 m/s^2.

6.1.1 Analysis of Horizontal Delivery Feed Track

In the track design for horizontal delivery (Fig. 6.1a), the last few parts cannot be fed, and even if the height of parts in the track is maintained at a satisfactory level, the delivery time will be greater than that given by Eq. (6.1).

Figure 6.2 shows the basic parameters defining the last portion of a horizontal delivery gravity feed track. This portion consists of a horizontal section AB of length L_1, preceded by a curved portion BC of constant radius R, which in turn is preceded by a straight portion inclined at an angle α to the horizontal. It is assumed in the following analysis that a certain fixed number of parts are maintained in the track above the delivery point. If the length of the straight inclined portion of the track containing parts is denoted by L_2, the number of parts N_p is given by

$$N_p = \frac{L_2 + R\alpha + L_1}{L} \qquad\qquad (6.2)$$

An equation is now derived giving the time t_p to deliver one part of length L. It is assumed in the analysis that the length of each part is small compared with the dimensions of the feed track and that the column of parts can be treated as a continuous, infinitely flexible rod.

When the escapement opens and the restraining force at A is removed, the column of parts starts to slide toward the workhead. In estimating the acceleration a of the column of parts, it is convenient to consider separately the parts in the three sections of the track, AB, BC, and CD, as shown in Fig. 6.3.

If the mass per unit length of the column of parts is denoted by m_1, the weight of section AB is given by $m_1 L_1 g$. The total frictional

Fig. 6.2 Idealized horizontal delivery gravity feed track.

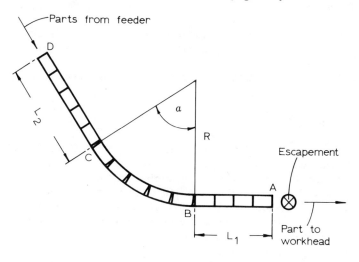

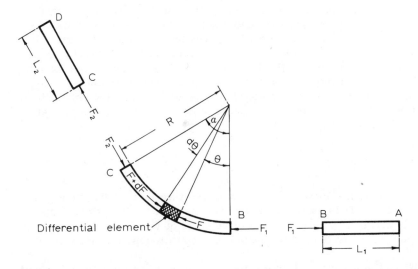

Fig. 6.3 Three separate sections of the idealized horizontal delivery track.

resistance in this region is given by $\mu_d m_1 L_1 g$ if μ_d is the coefficient of dynamic friction between the parts and the track.

The equation of motion for section AB is given by

$$F_1 = m_1 L_1 (\mu_d g + a) \qquad (6.3)$$

where F_1 is the force exerted on the parts in section AB by the remainder of the parts in the feed track and a is the initial acceleration of all the parts.

Similarly, the column of parts in the straight inclined portion CD of the feed track is partly restrained by a force F_2 given by

$$F_2 = m_1 L_2 (g \sin \alpha - \mu_d g \cos \alpha - a) \qquad (6.4)$$

To analyze the motion of the parts in the curved section BC of the feed track, it is necessary to consider an element of length $Rd\theta$ on a portion of the track that is inclined at an angle θ to the horizontal. In this case a force F is resisting the motion of the element and a force (F + dF) is tending to accelerate the element. These forces have a small component $F \, d\theta$ which increases the reaction between the parts and the track and consequently increases the frictional resistance.

The forces acting on the element are shown in Fig. 6.4 and the force equilibrium equation in the radial n direction is essentially

$$N_1 = m_1 gR \cos \theta \, d\theta + F \, d\theta \qquad (6.5)$$

The equation of motion in the tangential t direction is basically

$$dF + m_1 gR \sin \theta \, d\theta - \mu_d N_1 = m_1 Ra \, d\theta \qquad (6.6)$$

Substituting Eq. (6.5) in Eq. (6.6) and rearranging gives the first-order linear differential equation with constant coefficients

$$\frac{dF}{d\theta} - \mu_d F = m_1 R(a + \mu_d g \cos \theta - g \sin \theta) \qquad (6.7)$$

The general solution to Eq. (6.7) is

$$F = Ae^{\mu_d \theta} + m_1 gR \left[\frac{(1 - \mu_d^2) \cos \theta + 2\mu_d \sin \theta}{1 + \mu_d^2} - \frac{a}{\mu_d g} \right] \qquad (6.8)$$

Interestingly, there are two boundary conditions that are applicable: $\theta = 0$, $F = F_1$ and $\theta = \alpha$, $F = F_2$. The first boundary condition gives

$$A = F_1 - m_1 gR \left(\frac{1 - \mu_d^2}{1 + \mu_d^2} - \frac{a}{\mu_d g} \right) \qquad (6.9)$$

where F_1 is given by Eq. (6.3).

Fig. 6.4 Free body diagram for an element.

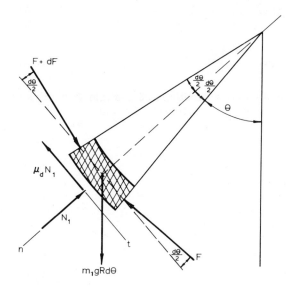

Equation (6.8) relates the acceleration of the line of parts along the curved and horizontal section when a force F is applied to the end of the line located at an angle θ. This applied force at the end is F_2 given by Eq. (6.4). Using this second boundary condition and solving Eq. (6.8) for a/g gives

$$\frac{a}{g} = \frac{\frac{L_2}{R}(\sin \alpha - \mu_d \cos \alpha) - \frac{L_1}{R}\mu_d e^{\mu_d \alpha} + \frac{(1 - \mu_d^2)(e^{\mu_d \alpha} - \cos \alpha) - 2\mu_d \sin \alpha}{1 + \mu_d^2}}{\frac{L_2}{R} + \frac{L_1}{R}e^{\mu_d \alpha} + \frac{e^{\mu_d \alpha} - 1}{\mu_d}}$$

(6.10)

Equation (6.10) shows the relationship between the nondimensional acceleration of the entire line a/g and the design parameters L_2/R, L_1/R, α, and μ_d. Since the part length is small compared to the total length of track, the acceleration is assumed constant, and thus the time t_p for one part to move into the workhead can be found from the kinematic expression

$$t_p = (2L/a)^{1/2}$$

(6.11)

where L is the length of the part along the track.

   Unfortunately, Eq. (6.10) is cumbersome to use for calculations and as a consequence, it is not easy to quickly determine the effect of a design change. It is also unrealistic to graph the information contained in Eq. (6.10) since there are five independent nondimensional design parameters. A nomogram would simplify this work, but Eq. (6.10) is not in a form suitable for this approach. However, Eq. (6.10) can be approximated by

$$\frac{a}{g} = f_1(\mu_d) + f_2(\frac{L_1}{R}) + f_3(\alpha) + f_4(\frac{L_2}{R})$$

(6.12)

which is a summation of independent functions of a single variable and is an acceptable form for a nomogram. The functions are

$$f_1 = 0.94(0.5 - \mu_d)$$

(6.13)

$$f_2 = -0.428(\frac{L_1}{R})$$

(6.14)

$$f_3 = 0.0084(\alpha - 45)$$

(6.15)

with α expressed in degrees, and

$$\frac{L_2}{R} = \frac{0.31e^{9.66(a/g)}}{1 + 0.018e^{-63(a/g)}} \tag{6.16}$$

Although no explicit function f_4 could be found to satisfy Eq. (6.12), there is a numerical relationship which can be developed from Eq. (6.16) for producing a nomogram.

Since Eq. (6.11) is also in a form suitable for nomographic presentation, the combined nomograms for Eqs. (6.10) and (6.11) are shown together in one nomogram in Fig. 6.5. This nomogram relates all the parameters that must be considered for the design of the lower section of a horizontal delivery gravity feed track.

6.1.2 Example

Suppose that a workhead is operating at a rate of 1 assembly per second and has 0.2 s to receive a 25 mm part; the value of L_1/R is ordinarily chosen as small as possible and in this situation equals zero; and the value of the dynamic coefficient of friction is 0.5. What are the values of L_2/R and α that will complete this design?

A line is drawn on the nomogram (Fig. 6.5) from a value of part length of 25 mm (1 in.) through a value of time of 0.2 s and intersects the a/g line at a value of 0.13. A second line is drawn from this value of a/g to a value of zero on the L_1/R scale. This line intersects turning line A. The third line, drawn from turning line A to a value of 0.5 on the μ scale, intersects turning line B. The last line is drawn from this intersection of turning line B through the α and the L_2/R scales. A number of solutions are possible; one such solution gives values of α and L_2/R of 60 degrees and 0.27, respectively. Substitution of these design parameters back into Eq. (6.10) yields a value for a/g of 0.12. The subsequent time from Eq. (6.11) is 0.207 s, which compared to the system design specification of 0.2 s is less than a 4% error.

Several comparisons were made to check the accuracy of this nomogram and the results showed that this nomogram was significantly accurate for design work. It is reasonable to expect the nondimensional acceleration found using this nomogram to be correct within 0.03. The worst results occur with small angles and large values of L_2/R or large coefficients of friction, but these situations are ordinarily avoided in practice.

It is interesting to note the effect of the value of the coefficient of friction on the value of the nondimensional acceleration a/g. From Eqs. (6.12) and (6.13) it is clear that an error in the value of the coefficient of friction produces a similar error in the value of a/g. Apparently, the source of the greatest error in this work is in the estimation of the value of the coefficient of friction. Some typical values are given in Table 6.1. A simple method for determining the dynamic coefficient of friction has also been developed that has the advantage

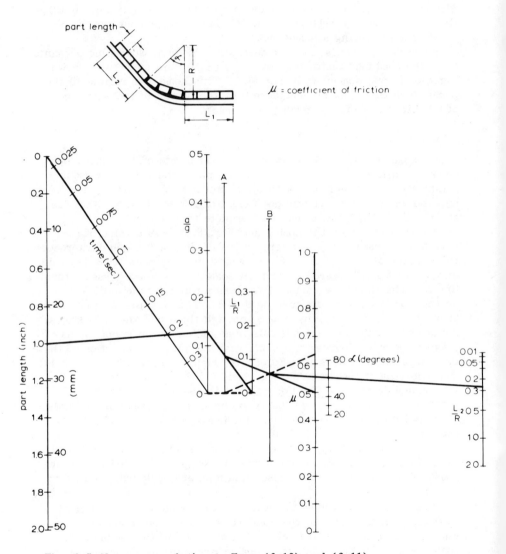

Fig. 6.5 Nomogram solution to Eqs. (6.10) and (6.11).

Table 6.1 Dynamic Coefficient of Friction of a Variety of Part
and Track Material Combinations

Part	Nylon	Plexiglass	Brass	Aluminum	Cast Iron	Steel
Nylon	0.520	0.536	0.568	0.475	0.375	0.503
Plexiglass	0.502	0.473	0.537	0.503	0.411	0.425
Brass	0.354	0.425	0.370	0.345	0.216	0.250
Aluminum	0.416	0.458	0.437	0.374	0.304	0.327
Cast Iron	0.314	0.370	0.368	0.268	0.218	0.252
Steel	0.349	0.419	0.432	0.353	0.273	0.306

(Track is a spanning header over: Nylon, Plexiglass, Brass, Aluminum, Cast Iron, Steel)

of using the actual parts in the experiments. The details are presented
in Appendix I.

It has tacitly been assumed that the static friction would not affect
this dynamic analysis. However, if the value of the static coefficient
of friction μ_s is significantly large, the parts in the queue simply will
not move. This would occur if the numerator in Eq. (6.10) were less
than or equal to zero when the value of the angle α was greater than
arctan μ_s. For this caluclation, the value of static coefficient of friction is substituted for the dynamic coefficient of friction μ_d. Using
the design parameters $L_1/R = 0$, $L_2/R = 0.27$, and $\alpha = 60$ degrees, the
nondimensional acceleration from Eq. (6.10) equals zero when the value
of the static coefficient of friction is 0.63. The dynamic analysis is
correct if the static coefficient of friction is less than this value.

Fortunately, the nomogram can also be used for this static analysis.
A line drawn between $a/g = 0$ and L_1/R intersects turning line A.
A line drawn from this intersection through the previous intersection
of turning line B crosses the μ line at the critical value of μ_s. Dashed
lines are used on Fig. 6.5 to show this procedure for the preceding
example.

6.1.3 On-Off Sensors

On automatic assembly machines it is essential to maintain a supply
of correctly oriented parts to the workhead. A condition when a workhead is prepared to carry out the assembly operation but is awaiting
the part to be assembled represents downtime for the whole assembly
line. Thus, for obvious economic reasons, correctly oriented parts
should always be available at each station on an assembly machine.
The probability that a part will be available at a workhead will be called
the reliability of feeding and should not be confused with the problems
associated with defective parts, which may jam in the feeding device

or workhead and then affect the reliability of the complete system. The latter subject is discussed in detail in Chapter 7.

When automatic feeding devices are used on assembly machines, the mean feed rate is usually set higher than the assembly rate to ensure a continuous supply of parts at the workhead. Under these circumstances, the feed tracks become full and parts back up to the feeder. For vibratory bowl feeders, however, this method is not always satisfactory. With these feeders, the parts in the line in the feed track are usually prevented from interfering with the orienting devices mounted on the bowl feeder track by a pressure brake. If a part filters past the orienting devices and arrives at the pressure break when the delivery chute is full, it is rejected back into the bowl. If this occurs too frequently, excessive wear of both the bowl and the parts can result.

To solve the problems caused by overfeeding and consequently ease the problem of wear of the bowl and parts, a simple control system can be employed which incorporates sensing devices located at two positions on the feed track. The feeder is turned ON when the level of parts falls below the sensing device nearest the workhead. The feeder is still set to deliver oriented parts at a higher rate than required and the parts gradually fill the delivery chute and back up to and then past the level of the first sensing device. When the second sensing device, which is located near the feeder, is activated, the bowl is turned OFF and the level of parts starts to fall again. When the level of parts falls below the level of the first sensing device, the first device is activated and the bowl feeder is again set in motion. Since the output of the bowl is a random variable, there may be some time delay before an oriented part leaves the bowl. Thus, it is possible that the entire line of parts between the lower sensor and the critical level found from Eq. (6.10) will be used by the workhead before an oriented part arrives from the feeder. This situation will now be studied theoretically to provide the designer of assembly machines with the information necessary to position the sensing devices in the feed track, so that the probability of the workhead being starved of parts is kept within acceptable limits.

Theory. Most vibratory bowl feeders are fitted with an orienting system which reorients or rejects those parts that would otherwise be fed incorrectly oriented. It is assumed that without the orienting system and for a particular setting, a vibratory bowl feeder would deliver n_f parts during the period of one workhead cycle. Thus, if η is the proportion of parts that pass through the orienting system, then $n_f \eta$ is the average number of parts that pass into the feed track during each workhead cycle. The factor η can be regarded as the efficiency of the orienting system as discussed in Chapter 5.

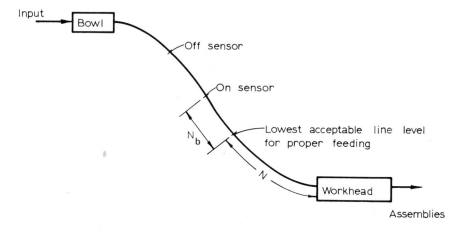

Fig. 6.6 Gravity feed track with On/Off sensors.

The output from the bowl feeder is assumed to have a binomial distribution with a mean of $n_f\eta$ and a variance of $n_f\eta(1-\eta)$ for each workhead cycle. For large values of n_f, a binomial distribution can be closely approximated by a normal distribution of the same mean and variance. The approximation improves as n_f increases but is always quite good if η is neither close to zero nor close to unity. Thus, the output from the feeder is approximately normal $(n_f\eta, n_f\eta(1-\eta))$ for every time span of one workhead cycle. The value of $n_f\eta$ must be greater than unity if the average value of the queue length is to increase. In ordinary steady-state queuing theory, $n_f\eta$ is termed the *traffic density* and is less than unity. For this reason, the results of steady-state queuing theory cannot be applied to this problem.

Referring to Fig. 6.6, N_b is the number of parts that can be held in the feed track between the lowest acceptable line level and the first sensing device. Thus, if t_w is the cycle time of the workhead, the distribution of parts that build up the line in $N_b t_w$ seconds is approximately normal $(N_b(n_f\eta-1), N_b n_f\eta(1-\eta))$, where N_b is subtracted from the mean to allow for the N_b parts that are assembled in $N_b t_w$ seconds.

For a particular value of N_b, Fig. 6.7 shows the density function for the net number of parts that build up the line in $N_b t_w$ seconds. The value of N_b corresponds to a particular reliability r, which means that the workhead will be starved of parts only $100(1-r)\%$ of the time following an activation of the ON sensor. The area under the density function in Fig. 6.7 equals r. An increase or decrease in N_b changes the mean $(N_b(n_f\eta-1))$ more than it changes the standard deviation

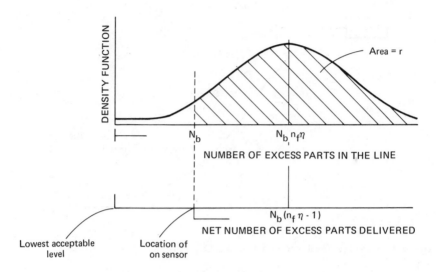

Fig. 6.7 Density functions for delivered parts. (From L. E. Murch and G. Boothroyd, "On-Off Control of Parts Feeding," *Automation*, vol. 18, no. 8, Aug. 1970, p. 32.)

$[N_b n_f \eta (1 - \eta)]^{1/2}$. Thus, the smaller the value of N_b, the smaller the area r under the density function and the lower the reliability.

Since the area under the curve in Fig. 6.7 is r, the average number of parts to leave the bowl feeder ($N_b n_f \eta$) is equal to n_s standard deviations $n_s[N_b n_f \eta (1 - \eta)]^{1/2}$, where n_s is determined from a normal distribution table. Thus,

$$N_b = \frac{n_s^2 (1 - \eta)}{n_f \eta} \tag{6.17}$$

For example, for a reliability of 0.99997 ($n_s = 4$),

$$N_b = \frac{16(1 - \eta)}{n_f \eta} \tag{6.18}$$

where $n_f \eta$ and η are known parameters for the bowl feeder.

In cases where these parameters are unknown, it may be convenient to let $N_b = n_s^2$, which is its maximum value since η is always positive and $n_f \eta$ must always be set greater than unity.

It has been stated that for a binomial distribution to be closely approximated by a normal distribution, n_f should be large. To determine a satisfactory value for n_f for particular values of η and r, a

Table 6.2 Simulation Results for Eq. (6.18)
(r = 0.99997)

η	n_f	$n_f \eta$
0.2	10	2
0.5	4	2
0.8	20	16

computer simulation routine was developed. Table 6.2 shows some of
the results from this simulation for r = 0.99997. For relatively large
values of η, the corresponding values of $n_f \eta$ may be too large to be
of any practical interest. However, when the theoretical value of N_b
was increased by 2, the same simulation yielded the results shown in
Table 6.3. Thus, for $n_f \eta \geq 1.6$ (r = 0.99997),

$$N_b = \frac{16(1 - \eta)}{n_f \eta} + 2 \tag{6.19}$$

Equation (6.19) can be used to estimate the number of parts between
the lower sensor and the level of parts for acceptable operation when
the feeder is capable of delivering at a rate 60% greater than the workhead assembly rate, a situation that is not uncommon when on-off sensors are employed. An examination of Eq. (6.19) indicates that the
maximum value of N_b is 12. This occurs when η is small and $n_f \eta$ equals
1.6, its minimum. Other conditions that satisfy the constraint on Eq.
(6.19) will always produce results for N_b less than or equal to 12.
Thus, when little is known about a feeder except that it significantly
overfeeds, the lower sensor can be placed in a position where N_b equals
12 and satisfactory performance is assured.

The number of parts that can be held between sensing devices has
no effect on the average running time of the feeder but does effect
the average frequency of activations of the sensing devices.

Table 6.3 Simulation Results for Eq. (6.19)
(r = 0.99997)

η	n_f	$n_f \eta$
0.1	16	1.6
0.2	8	1.6
0.5	3	1.5
0.8	2	1.6

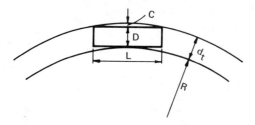

Fig. 6.8 Construction to determine minimum diameter of a curved feed track.

Often, only one sensor is used. Besides activating the feeder it activates a time delay to simulate the second sensor. Several types of sensors, including fiber optics or pneumatics, can be used on the track, but the basic logic circuitry governing their operation is identical.

6.1.4 Feed Track Section

A compromise is necessary when designing the feed track section. The clearances between the part and the track must be sufficiently large to allow transfer and yet must be small enough to keep the part from losing its orientation during transfer. In the curved portions of the track further allowances have to be made to prevent the part jamming. Figure 6.8 shows a cylindrical part in a curved tubular track. For the part to negotiate the bend, the minimum track diameter d_t is given by

$$d_t = c + D \tag{6.20}$$

and by geometry

$$(R + d_t - c)^2 + (\frac{L}{2})^2 = (R + d_t)^2$$

or

$$2c(R + d_t) - c^2 = (\frac{L}{2})^2 \tag{6.21}$$

where R is the inside radius of the curved track and L the length of the part. If c is small compared with $2(R + d_t)$, this expression becomes approximately

$$2c(R + d_t) = (\frac{L}{2})^2 \qquad\qquad (6.22)$$

Substituting for c from Eq. (6.20) into Eq. (6.22) and rearranging gives

$$d_t = 0.5\{[(R + D)^2 + \frac{L^2}{2}]^{1/2} - (R - D)\} \qquad\qquad (6.23)$$

If the parts are sufficiently bent or bowed, it may be difficult to design a curved track that will not allow overlapping of parts and consequent jamming. This is illustrated in Fig. 6.9.

Figure 6.10 illustrates typical track sections used for transferring cylinders, flat plates, and headed parts. An important point to be remembered when designing a feed track is that the effective coefficient of friction between the parts and the track may be higher than the actual coefficient of friction between the two materials. Figure 6.11 gives some examples of the effect of track cross section on the effective coefficient of friction. The increase of 100% in friction given by the example in Fig. 6.11d could have very serious consequences in a gravity feed system. It is also important to design these tracks with removable covers or access holes for the quick removal of jammed parts.

6.1.5 Design of Gravity Feed Tracks for Headed Parts

Of the many parts that can be fed in a gravity feed track, perhaps the most common are headed parts such as screws and rivets, which are often fed in the manner shown in Fig. 6.12. Clearly, if the track inclination is too small or if the clearance above the head is too small, the parts will not slide down the track. It is not always realized, however, that the parts may not feed satisfactorily if the track has

Fig. 6.9 Blockage in a curved feed track

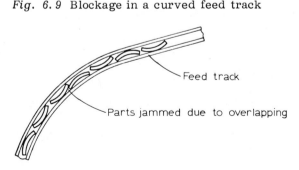

Feed track

Parts jammed due to overlapping

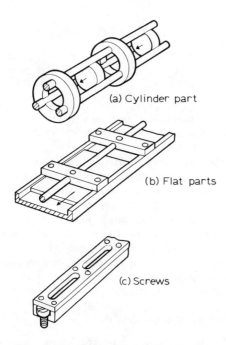

(a) Cylinder part

(b) Flat parts

(c) Screws

Fig. 6.10 Various gravity feed track sections for typical parts.

Fig. 6.11 Relation between effective coefficient of friction μ and actual coefficient of friction μ_d for various track designs.

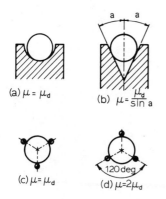

(a) $\mu = \mu_d$

(b) $\mu = \dfrac{\mu_d}{\sin a}$

(c) $\mu = \mu_d$

(d) $\mu = 2\mu_d$

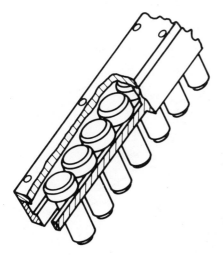

Fig. 6.12 Headed parts in a gravity feed track.

too steep an inclination or if too large a clearance is provided between
the head of the part and the track. Also, under certain circumstances
a part may not feed satisfactorily, whatever the inclination or clearance.
This analysis of the design of gravity feed tracks for headed parts
provides the designer of assembly machines and feeding devices with
the information necessary to avoid situations where difficulty in feeding
will occur.

   Analysis.   Figure 6.13 shows a typical headed part (a cap screw
with a hexagon socket) in a feed track. It is clear that as the track
inclination θ is gradually increased and provided that the corner B of
the screw head has not contacted the lower surface of the cover of
the track, the screw will slide when $\theta > \arctan \mu_1$, where μ_1 is a function of the coefficient of static friction between the screw and the track.
On further increases in the track inclination, the condition shown in
Fig. 6.14 will eventually arise when the corner B of the screw head has
just made contact with the lower surface of the cover of the track.
Immediately prior to this condition, the center of mass of the screw lies
directly below AA, a line joining the points of contact between the
screw head and the track. From Fig. 6.14 it can be seen that z, the
distance from the line AA to the axis of the screw, is given by

$$z^2 = \frac{d^2 - s^2}{4}$$

(6.24)

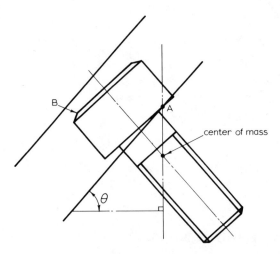

Fig. 6.13 Position of a headed part that does not touch the track cover.

Fig. 6.14 Position of a headed part that when corner B just contacts the lower surface of the track cover.

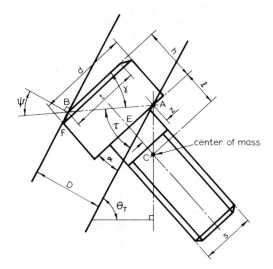

where s is the width of the slot and the diameter of the shank and d is the diameter of the screw head.

Also from the triangle ACE (Fig. 6.14)

$$\frac{z}{\ell} = \tan (\theta_T - \alpha) \qquad (6.25)$$

where ℓ is the distance from the center of mass of the screw to a plane containing the underside of the head, α is the angle between the screw axis and a line normal to the track, and θ_T, which will be called the tilt angle, is the track angle at which the top of the screw head just contacts the lower surface of the top cover of the track. From Fig. 6.14,

$$\alpha = \tau - \gamma \qquad (6.26)$$

where

$$\tau = \arcsin \left\{ \frac{D}{[h^2 + (t/2 + z)^2]^{1/2}} \right\} \qquad (6.27)$$

and

$$\gamma = \arctan \left(\frac{h}{t/2 + z} \right) \qquad (6.28)$$

where D is the depth of the track, h the depth of the screw head, and t the diameter of the top of the screw head. Thus, combining Eqs. (6.25) through (6.28) gives

$$\theta_T = \arctan \left(\frac{z}{\ell} \right) - \arctan \left(\frac{h}{t/2 + z} \right) + \arcsin \left\{ \frac{D}{[h^2 + (t/2 + z)^2]^{1/2}} \right\} \qquad (6.29)$$

Sliding will always occur if $\theta_T > \theta > \arctan \mu_1$, since under these circumstances, the track angle is greater than the angle of friction and there is no contact between the top of the screw head and the track cover. However, sliding may still occur if θ is larger than θ_T. This situation is shown in Fig. 6.15, where it can be seen that a frictional force occurs between the screw head and the track cover as well as on the lower portion of the track.

For sliding to occur under these conditions

$$mg \sin \theta > F_1 + F_2 \qquad (6.30)$$

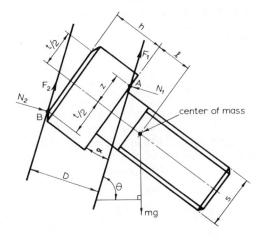

Fig. 6.15 Forces acting on a cap screw in a feed track.

where

$$F_1 = \mu_1 N_1$$

and
(6.31)

$$F_2 = \mu_2 N_2$$

where μ_1 and μ_2 are the effective values of the coefficient of static friction between the part and the track at A and B, respectively (Fig. 6.15). Resolving forces normal to the track gives

$$N_1 = N_2 + mg \cos \theta \qquad (6.32)$$

and taking moments about A gives

$$mg[\ell \sin (\theta - \alpha) - z \cos (\theta - \alpha)]$$
$$= N_2 [(\frac{t}{2} + z) \cos \alpha - h \sin \alpha] - F_2 D \qquad (6.33)$$

Substituting Eqs. (6.31) through (6.33) in Eq. (6.30) and rearranging gives

$$\sin \theta > \mu_1 \cos \theta + (\mu_1 + \mu_2) \frac{\ell \sin (\theta - \alpha) - z \cos (\theta - \alpha)}{[(t/2) + z] \cos \alpha - h \sin \alpha - \mu_2 D}$$
(6.34)

where μ_1 is the effective coefficient of friction between the screw head and the bottom portion of the track and μ_2 is the effective coefficient of friction between the screw head and the track cover.

Under the conditions shown in Figs. 6.13 through 6.15, a portion of the screw head lies below the lower contact surface of the track and the effective coefficient of static friction between the screw and the track is greater than the actual coefficient of static friction μ_S for the two materials. It can be shown that, for this situation, the relationship between μ_S and μ_1 is given by

$$\mu_1 = \mu_S \left[\left(\frac{d \sin \alpha}{2z} \right)^2 + \cos^2 \alpha \right]^{1/2} \qquad (6.35)$$

In this particular case, the difference between μ_S and μ_1 is small, but in some cases discussed later in this section, this effect is of importance. Since there is point contact between the track cover and the screw head, $\mu_2 = \mu_S$.

Thus, there are at least two conditions governing the motion of the screw in the track. First, if the track angle is greater than the angle of friction and less than or equal to the tilt angle θ_T, the screw slides. Under these circumstances, the greater the tilt angle θ_T, the larger the range of values of μ_S for which the screw slides and, therefore, as can be seen from Eq. (6.29), the value of D, the track depth, should be as large as possible.

Second, the maximum track angle is restricted to the value given by Eq. (6.34) and in this case, again, the greater the depth of the track, the greater will be this maximum value of θ. It is of interest now to determine the critical value of track depth below which a screw cannot jam in the track, since this will allow a complete definition of the range of track angles and track depths for which a part will feed satisfactorily.

If the track depth D is gradually increased, a special case of the condition shown in Fig. 6.15 will eventually arise. This is where the angle between a line joining A and B and a line normal to the track becomes equal to the angle of friction between the screw and the track. For larger values of D, the part will not normally make contact with the upper portion of the track. However, under these circumstances the screw will jam in the track if it makes contact across A and B. Thus, the situation arises where, theoretically, the screw should slide, but if a small perturbation rotates it sufficiently to contact the cover, it would lock in the track. This possibility should clearly be avoided in practice. From the geometry of Fig. 6.15, the maximum value of D is thus given by

$$D_{max} = \left[h^2 + \left(\frac{t}{2} + z \right)^2 \cos \beta_2 \right]^{1/2} \qquad (6.36)$$

where $\tan \beta_2 = \mu_2 = \mu_s$.

Figure 6.16 shows a graph of θ plotted versus D/h for a particular screw where $d = 1.712s$, $h = s$, $\ell = 1.5s$, and $t = 1.42s$. These screw dimensions were taken from "Machinery's Handbook" and are the American Standard for a size 8 hexagon socket-type cap screw.

Although in practice the width of the track s would be larger than the diameter of the screw shank, it should be as small as possible and in all the examples dealt with in this section, it is considered to be equal to the shank diameter.

The results in the figure show the ranges of values of θ and D/h for which a screw will slide down the track without the possibility of jamming for various values of μ_s. It can be seen that, as the coefficient of friction is increased, the ranges of value of θ and D for feeding to occur decrease. It should also be noted that the line XX, representing the tilt angle θ_T, passes through the points that give (1) the minimum track angle and minimum track depth, and (2) the maximum track angle and maximum track depth.

It is now suggested that a reasonable criterion for the best track inclination and track depth would be one where these parameters are such that feeding would occur for the widest range of values of μ_s. This condition is where the points defined under (1) and (2) above become identical. Referring to Fig. 6.16, this condition will occur when $\theta_{max} = \theta_{min}$ and $D_{max} = D_{min}$, that is, when $\theta = 39$ degrees and $D/h = 1.32$, and for this situation it can be seen that the screw will always slide if $\mu_s < 0.81$.

Now θ_{min} occurs when $\theta_T = \beta_1$, where $\tan \beta_1 = \mu_1$. Substituting in Eq. (6.25) gives

$$\tan (\beta_1 - \alpha) = \frac{z}{\ell}$$

or

$$\beta_1 - \alpha = \arctan \left(\frac{z}{\ell}\right) \tag{6.37}$$

When θ equals θ_{max}, it also equals θ_T; hence from Fig. 6.14, the value of θ_{max} occurs when ψ equals β_2, which gives

$$\tan (\alpha + \beta_2) = \frac{z + t/2}{h}$$

or

$$\alpha + \beta_2 = \arctan \left(\frac{z + t/2}{h}\right) \tag{6.38}$$

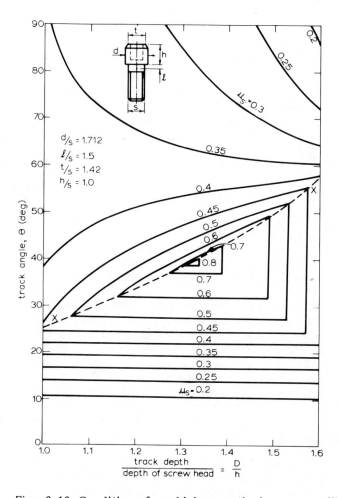

Fig. 6.16 Conditions for which a particular screw will slide in a gravity feed track. (From A. H. Redford and G. Boothroyd, "Designing Gravity Feed Tracks for Headed Parts," Automation, vol. 17, no. 5, May 1970, p. 96.)

The relationships among β_1, β_2, and β are given by

$$\tan \beta_1 = \tan \left[\left(\frac{d \sin \alpha}{2z} \right)^2 + \cos^2 \alpha \right] \tag{6.39}$$

and

$$\beta_2 = \beta \tag{6.40}$$

Solving Eqs. (6.37) through (6.40) simultaneously will give the maximum angle of friction β for which the screw will slide without the possibility of jamming. Substitution of this value in Eqs. (6.39) and (6.36) will give the corresponding optimum values of the track inclination $\theta (= \beta_1)$ and the track depth D.

Table 6.4 gives the equations necessary to determine the maximum coefficient of friction μ_{max} (= tan β), the track inclination, and the track depth for the proposed optimum conditions for four common types of screw:

1. A cap screw of hexagon socket type
2. A flat head cap or machine screw in a V track
3. A flat head cap or machine screw in a plain track
4. A button head cap or round head machine screw

In all cases it has been assumed that the width of the slot in the track is equal to the diameter of the screw shank. The equations listed in Table 6.4 were derived by considering the geometry of the part in the track at the condition where the coefficient of friction is a maximum. Referring to Fig. 6.14, this condition occurs for the cap screw when the center of mass of the screw lies directly below A, when the effective angle of friction at A equals the track angle θ_T, and when the angle between the line AB and the normal to the track at B equals the effective angle of friction at B (that is, when $\psi = \beta$). Figures 6.17 through 6.19 show this critical condition and the notation for the other three types of screw analyzed.

The expressions presented in Table 6.4 for the cap head screw of the hexagon socket type and the flat head cap screw in a plain track are valid only under certain circumstances. The analysis for the former is valid provided that point F in Fig. 6.14 does not contact the underside of the upper part of the track, and the analysis for the latter is valid provided that point A in Fig. 6.18 is below the upper surface of the screw head. Simple analyses not presented here have shown that the equations derived are valid for the range of conditions presented.

Table 6.4 Equations Necessary to Determine the Maximum Coefficient of Friction, μ_{max} (= tan β), the Track Inclination θ, and the Track Depth D for Optimum Feeding Conditions

---

(1) Cap head screw of the hexagon socket type

z^2 $\qquad\qquad = (d^2 - s^2)/4$

$\tan \theta$ $\qquad\qquad = \tan \beta [(d \sin \alpha / 2z)^2 + \cos^2 \alpha]^{1/2}$

$\tan (\theta - \alpha)$ $\qquad = z/\ell$

$\tan (\theta + \beta)$ $\qquad = [\ell h + z/(z + t/2)]/[\ell/(z + t/2) - z/h]$

D $\qquad\qquad\quad = h \cos \beta \{1 + [(z + t/2)/h]^2\}^{1/2}$

(2) Flat head cap or machine screw in a V track

$\tan \phi$ $\qquad\qquad = (d - s)/2h$

$d/2\ell$ $\qquad\qquad = (\tan^2 \beta - \sin \phi)(\cos^2 \beta + \cot^2 \phi)^{1/2}/\sin \beta (1 + \sin \phi)$

D $\qquad\qquad\quad = d \cos \beta [1/2(1 + \cos \beta)/(\cos^2 \beta + \cot^2 \phi)^{1/2}]$

$\qquad\qquad\qquad + d \cot^2 \phi / [2(\cos^2 \beta + \cot^2 \phi)^{1/2}] - (s \cot \phi)/2$

$\tan \theta$ $\qquad\qquad = \tan \beta / \sin \phi$

(3) Flat head cap or machine screw in a plain track

$\tan \theta$ $\qquad\qquad = (d - s)/2h$

$y \tan \phi + (s/2)$ $\quad = [z^2 + (s/2)^2]^{1/2}$

$\tan \alpha$ $\qquad\qquad = z/\{\tan \phi [z^2 + (s/2)^2]^{1/2}\}$

$z/(y + \ell - h)$ $\qquad = \tan (\theta - \alpha)$

$\tan \gamma$ $\qquad\qquad = 2 \sec^2 \phi (q \sin \alpha \cos \alpha + w \cos^2 \alpha - w \cos^2 \phi)/s$

$\tan \theta$ $\qquad\qquad = \tan \beta / \sin \gamma$

$\tan (\beta + \alpha)$ $\qquad = (z + d/2)/(h - y)$

q $\qquad\qquad\quad = \sin \alpha [z^2 + (s/2)^2]^{1/2}/\tan \phi + z \cos \alpha$

w $\qquad\qquad\quad = \cos \alpha [z^2 + (s/2)^2]^{1/2}/\tan \phi - z \sin \alpha$

(4) Button head cap or round head machine screws

z^2 $\qquad\qquad\quad = (d^2 - s^2)/4$

$\tan \theta$ $\qquad\qquad = \tan \beta \{[(d \sin \alpha)/2z]^2 + \cos^2 \alpha\}^{1/2}$

$\tan (\theta - \alpha)$ $\qquad = z/\ell$

$\beta + \gamma + \alpha$ $\qquad = 90 \text{ degrees}$

$\tan \gamma$ $\qquad\qquad = [h + T(\cos \alpha - 1)]/(z + T \sin \alpha)$

T $\qquad\qquad\quad = [(d/2)^2 + h^2]/2h$

D $\qquad\qquad\quad = z \sin \alpha + h \cos \alpha - T(\cos \alpha - 1)$

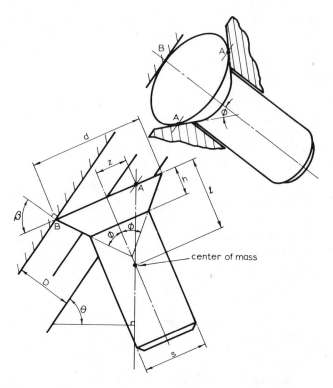

Fig. 6.17 Flat head cap or machine screw in V track.

Results. Graphs of μ_{max}, D/h, and θ versus ℓ/s for the four types of screw studied are shown in Figs. 6.20 through 6.23. Each figure shows these values for the two extreme geometries of the particular type of screw. For example, in the case of a cap head screw of the hexagon socket type, regardless of its size, all values of d/s and t/s fall within the ranges 1.33 to 1.712 and 1.041 to 1.42, respectively.

Clearly, the value of D/h cannot be less than unity, and thus for some types of screw, it can be seen from the figures that there is a minimum value of ℓ/s for which the analysis is valid. If ℓ/s is less than this minimum value, the parts can be fed in a track with a small clearance and the only restriction on the track angle is that it must be greater than the effective angle of friction.

Although in practice there will often be no choice in the type of screw to be used, it is of interest to note the degree of difficulty in feeding equivalent sizes of the various screws analyzed. For a given

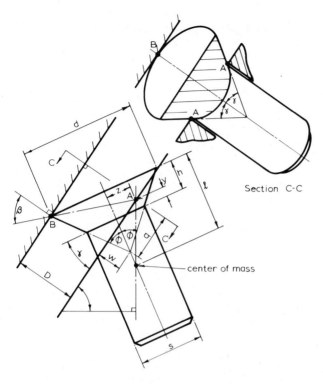

Section C-C

Fig. 6.18 Flat head cap or machine screw in plain track.

Fig. 6.19 Button head cap or round head machine screw in plain track.

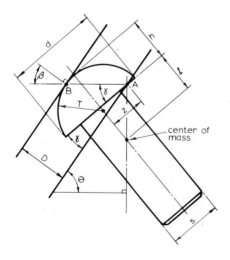

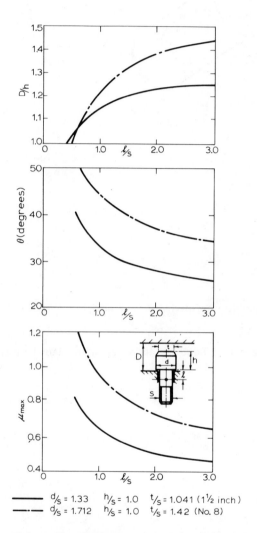

Fig. 6.20 Optimum values of track angle θ and track depth D for maximum coefficient of friction μ_{max}: cap head or hexagon socket screws. (From A. H. Redford and G. Boothroyd, "Designing Gravity Feed Tracks for Headed Parts," Automation, vol. 17, no. 5, May 1970, p. 96.)

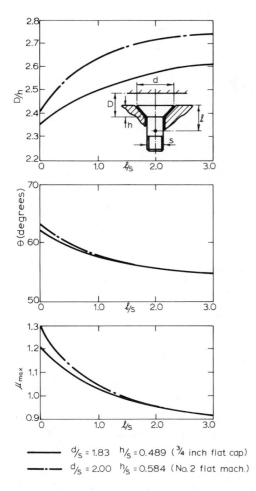

Fig. 6.21 Optimum values of track angle θ and track depth D for maximum coefficient of friction μ_{max}: flat head cap or machine screws in V track. (From A. H. Redford and G. Boothroyd, "Designing Gravity Feed Tracks for Headed Parts," Automation, vol. 17, no. 5, May 1970, p. 96.)

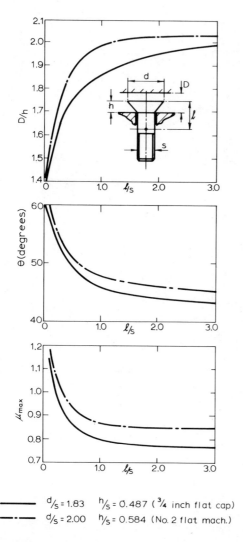

- ——— $d/s = 1.83$ $h/s = 0.487$ (¾ inch flat cap)
- —·— $d/s = 2.00$ $h/s = 0.584$ (No. 2 flat mach.)

Fig. 6.22 Optimal values of track angle θ and track depth D for maximum coefficient of friction μ_{max}: flat head cap or machine screws in plain track. (From A. H. Redford and G. Boothroyd, "Designing Gravity Feed Tracks for Headed Parts," Automation, vol. 17, no. 5, May 1970, p. 96.)

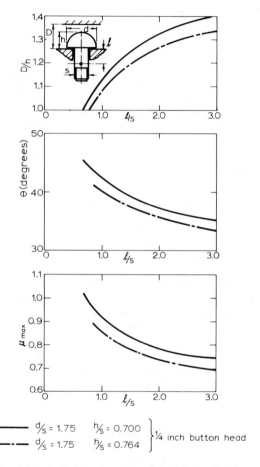

Fig. 6.23 Optimal value of track angle θ and track depth D for maximum coefficient of friction μ_{max}: button head cap or round head machine screws. (From A. H. Redford and G. Boothroyd, "Designing Gravity Feed Tracks for Headed Parts," Automation, vol. 17, no. 5, May 1970, p. 96.)

length of screw, the most difficult to feed are the larger sizes of cap head screws of the hexagon socket type, and in this case if μ_S is 0.5, it would be very difficult to feed the screw if its length were greater than five times its diameter. The easiest screws to feed are the flat head cap or machine screws and of the two alternative track designs examined, the V track has better feeding characteristics.

One general point is that, in many cases, the optimum track depth is very much larger than would normally be used in practice, and when the track depth is only slightly larger than the depth of the head, difficulty is often encountered when feeding these parts.

Procedure for Use of Figs. 6.20 Through 6.23. The first step in using the data provided in the figures is to determine the value of ℓ/s for the screw under consideration. The simplest way would be to balance the screw on a knife edge and measure the distance ℓ from the center of mass to the underside or top of the screw head, whichever is appropriate. This value would then be divided by the shank diameter to give the ratio ℓ/s.

The optimum values of the track angle θ and the track depth to screw head depth ratio are then read off the appropriate curves in the figures. In all cases, except the cap head screw of hexagon socket type (Fig. 6.20), the ranges of these values are quite small. When using this figure, therefore, it should be remembered that the lower line is for the large sizes and the upper line is for the small sizes.

Finally, reference to the lower graph will give the maximum permissible value of μ_S. If the actual value of μ_S is greater than this, the screw cannot be fed in a slotted gravity feed track.

Although the gravity feed track is the simplest form of feed track, it has some disadvantages. The main disadvantage is the need to have the feeder in an elevated position. This may cause trouble in loading the feeder and in freeing any blockages that may occur. In such cases, the use of a powered track may be considered.

6.2 Powered Feed Tracks

The most common types of powered feed tracks are vibratory tracks and air-assisted tracks. A vibratory feed track is illustrated in Fig. 6.24 and operates on the same principle as the vibratory bowl feeder (Chapter 3). With this device, the track is generally horizontal and its performance is subject to many of the limitations of a conventional vibratory bowl feeder. In the feeder shown, the vibrations normal and parallel to the track are in phase and the feeding characteristics will be affected by changes in the effective coefficient of friction μ between the parts and the track. This is illustrated in Fig. 3.8, where

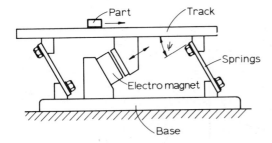

Fig. 6.24 Horizontal vibratory feed track.

it can be seen that an increase in μ generally gives an increase in the conveying velocity. For example, a typical operating condition would be where the normal track acceleration, $A_n = 1.1g_n$. In this case, stable feeding would occur and a change in μ from 0.2 to 0.8 would give an increase in conveying velocity from 18 to 55 mm/s when operating at a frequency of 60 Hz. It is shown in Appendix II that introducing the appropriate phase difference between the components of vibration normal and parallel to the track can give conditions where the conveying velocity is consistently high for a wide range of values of μ. With a track inclined at 4 degrees (0.07 rad) to the horizontal (a typical figure for a bowl feeder), the optimum phase angle for a normal track acceleration of 1.2g is approximately -65 degrees (1.1 rad). Figure II.2 shows that this figure is not significantly affected by the vibration angle employed. In the work that led to these results it was also found that the optimum phase angle reduces as the track angle is reduced until, when the track is inclined downward at an angle of 8 degrees (0.14 rad), the optimum phase angle is almost zero. This means that a drive of the type shown in Fig. 6.24 operates under almost optimum conditions for a track inclined at 8 degrees downward. Under these circumstances it is found that the mean conveying velocity for a wide range of μ will be given by

$$v_m = \frac{4500}{f} \quad mm/s \qquad (6.41)$$

where f is the frequency of vibration in hertz when the vibration angle is 20 degrees (0.35 rad) and A_n/g_n is 1.2. In this case, the mean conveying velocity for an operating frequency of 60 Hz would be 75 mm/s. A higher feed rate could be obtained by reducing the vibration angle, increasing the downward slope of the track, or reducing the frequency of vibration.

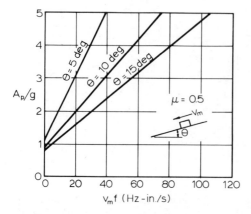

Fig. 6.25 Effect of parallel track acceleration on mean conveying velocity.

For downward-sloping tracks with a large inclination, large feed rates can be obtained with zero vibration angle. In this case the track is simply vibrated parallel to itself; typical feed characteristics thus obtained are illustrated in Fig. 6.25, which shows the effect of parallel track acceleration on the mean conveying velocity for various track angles when the coefficient of friction is 0.5. The results for a feed track with parallel vibration can be summarized by the empirical equation

$$v_m = \frac{(\theta/f)(A_p/g + 9.25 \times 10^{-4} \theta^2 - 2.3\mu + 0.25)(25.4)}{0.007\theta + 0.07 + 0.8\mu} \qquad (6.42)$$

where

v_m = mean conveying velocity, mm/s
f = vibration frequency, Hz
θ = inclination of downward sloping track, degrees
μ = effective coefficient of friction between part and track
A_p = parallel track acceleration, mm/s^2
g = acceleration due to gravity (9810 mm/s^2)

Equation (6.42) applies when $A_p/g \geq 2.0$, $0.3 \leq \mu \leq 0.8$, $5 \leq \theta \leq 25$ and when $\theta \leq 28.5\mu + 2.5$.

6.2.1 Example

A track inclined downward at 10 degrees is vibrated parallel to itself at a frequency of 60 Hz with an amplitude of vibration of 0.2 mm. If the coefficient of friction between the part and the track is 0.5, the mean conveying velocity of the part may be estimated as follows. The dimensionless parallel track acceleration

$$\frac{A_p}{g} = \frac{(0.2)(120\pi)^2}{9810} = 2.9$$

Since $\theta \le 28.5\mu + 2.5$ and $A_p/g > 2.0$, Eq. (6.42) applies and therefore

$$v_m = \frac{(10/60)[2.9 + 9.25 \times 10^{-4} \times 10^2 - 2.3(0.5) + 0.25](25.4)}{0.007(10) + 0.07 + 0.8(0.5)} = 16 \text{ mm/s}$$

Air-assisted feed tracks are often simply gravity feed tracks with air jets suitably placed to assist the transfer of the parts (Fig. 6.26). These devices are ideal for conditions where the gravity feed track alone will not quite meet the requirements. Although a well-designed air-assisted feed track can, under suitable conditions, feed parts up an inclined track, it is more usual for the track to slope downward or to be horizontal.

For all types of feed tracks there are two important points that must be considered. For good space utilization a feed track should not be too long. Conversely, the feed track, besides acting as a transfer device, provides a buffer stock of parts which, if a blockage occurs, will allow the workhead to continue operating for a limited period. Ideally, the feed track should be capable of holding enough parts to ensure that the workhead can continue to operate for a time sufficient to allow the blockage to be detected and cleared.

Fig. 6.26 Air-assisted feed track.

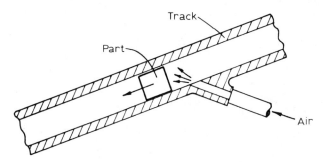

A further requirement for all feed tracks, and indeed all parts feeders and workheads, is that in the event of a blockage, the parts are readily accessible. For this reason, feed tracks should be designed to allow easy access to all parts of the track.

6.3 Escapements

Many types of escapement have been developed and quite often, for a given part, there are several different types available that will perform the required function. Figure 6.27 shows two examples of probably the simplest type of escapement. Here the parts are pulled from the feed track by the work carrier and the escapement itself consists of only a rocker arm or a spring blade.

Many escapements are not always recognized as such. For example, a rotary indexing table may be arranged to act as an escapement. This is shown in Fig. 6.28, where parts may be taken from either a horizontal delivery feed track or a vertical delivery feed track.

An advantage of the simple escapements illustrated in Figs. 6.27 and 6.28 is that they also act as parts placing mechanisms. The two escapements shown in Fig. 6.27 are somewhat unusual because the escapement is activated by the work carrier and part, whereas with most escapements, the motion of the part is activated by the escapement, which is in turn activated by some workhead function. This latter method is the most common in practice and escapements of this type may be subdivided into various categories, as described below.

Fig. 6.27 Escapements actuated by the work carrier.

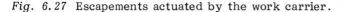

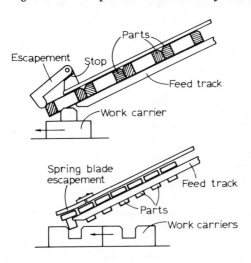

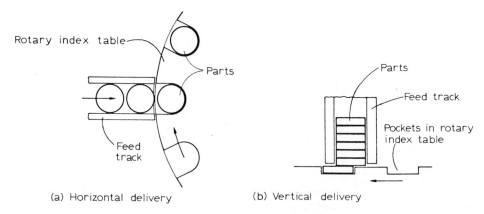

(a) Horizontal delivery   (b) Vertical delivery

Fig. 6.28 Feeding of parts onto rotary index table.

6.3.1 Ratchet Escapements

Two examples of ratchet escapements are shown in Fig. 6.29; the functions being performed are different in each case. The pawl is designed so that as its front finger lifts clear of the line of parts, its back finger retains either the next part, as shown in Fig. 6.29a, or a part further up the line, as shown in Fig. 6.29b. Ratchet escapements operating on several feed tracks can be activated from a single mechanism. In automatic assembly, the release of several parts from a single feed track is not often required, but the release of one part from each of several feed tracks is often desirable. This can be achieved by a series of ratchet escapements of the type shown in Fig. 6.29a. As the escapement is activated, the front finger rotates, but without moving the part to be released. At position 2, the front finger is just about to release part A and the back finger moves in such a way that no motion is imparted to part B. On the return stroke of the escapement, part B is released by the back finger and retained by the front finger. Figure 6.29b shows a similar type of mechanism except that the remaining parts in that track move forward before being retained by the back finger. In these devices the back finger of the escapement should not produce motion of the parts opposite to the direction of flow. If this tendency is present, either all the parts above this point on the feed track will move backward and may subject the escapement to heavy loads or the parts may lock and cause damage to the escapement.

In the examples already shown, the motions of the front and back fingers of the escapement are obtained by a rotary motion. The fingers of a ratchet escapement may, however, be operated together

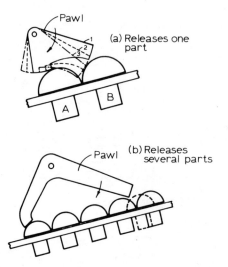

Fig. 6.29 Ratchet escapements operated by rotary motion.

or independently by cams, solenoids, or pneumatic cylinders, giving a linear motion as shown in Fig. 6.30.

It is clear from all the foregoing examples of ratchet escapements that the escapement can only be used to regulate the flow of parts which, when arranged in single file, have suitable gaps between their outer edges.

Fig. 6.30 Ratchet escapements operated by linear motion.

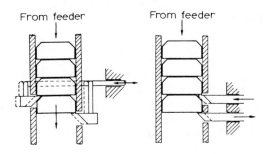

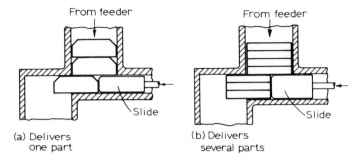

(a) Delivers
    one part

(b) Delivers
    several parts

Fig. 6.31 Slide escapements delivering into single feed chute.

6.3.2 Slide Escapements

Five examples of slide escapements are shown in Figs. 6.31 through 6.33. It can be seen from the figures that in the slide escapement, one or more parts are removed from the feed chute by the action of a cross-slide and that applications of this type of device are restricted to parts that do not interlock with each other. The slide escapement is ideally suited to regulating the flow of spherical, cylindrical, or platelike parts and although in all the figures the feed track enters the escapement vertically, this, although desirable, is not necessary.

As with the ratchet escapement, parts may be released either singly or in batches from one or a number of feed tracks by the action of a single actuating mechanism. However, a further alternative, not available with the ratchet escapement, is for parts fed from a single feed track to be equally divided between two delivery tracks, as shown in Fig. 6.33a. This type of escapement is very useful where two identical

Fig. 6.32 Slide escapements operating several feed chutes.

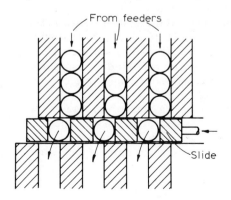

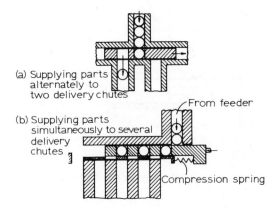

(a) Supplying parts alternately to two delivery chutes

(b) Supplying parts simultaneously to several delivery chutes

From feeder

Compression spring

Fig. 6.33 Slide escapements supplying two or more delivery chutes from a single feed chute.

parts are to be used in equal quantities and a parts feeder is available that will deliver at a sufficient rate to meet the total requirement. If it is necessary to feed parts from a single feed track into more than two delivery tracks, a slide escapement of the type shown in Fig. 6.33b would be suitable. The figure shows three delivery tracks being fed from a single feed track and, as before, only one actuator is necessary.

6.3.3 Drum Escapements

Two types of drum escapement, usually referred to as drum-spider escapements, are shown in Fig. 6.34, where in these cases the drum is mounted vertically and the parts are either fed and delivered side by side (Fig. 6.34a) or fed end to end and delivered side by side (Fig. 6.34b). In the latter case the parts are fed horizontally to the escapement. One advantage of the vertical drum escapement is that a change in the direction of motion of the parts is easily accomplished. This can be very useful where the horizontal distance between the parts feeder and the workhead is restricted to a value that would necessitate very sharp curves in the feed track if an alternative type of escapement were used. A further feature of the drum-spider escapement is that on passing through the escapement, the parts can be turned through a given angle.

Two other types of drum escapement, the star-wheel and the worm, are shown in Fig. 6.35, where it can be seen that the direction of motion of the parts is unaffected by the escapement and it is necessary that the parts have a suitable gap between their outer edges when they are arranged in single file.

(a) Parts fed and
    delivered side
    by side

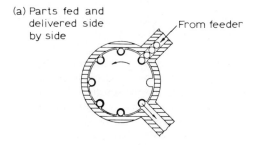

From feeder

(b) Parts fed end to end and
    delivered side by side

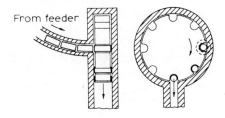

From feeder

Fig. 6.34 Drum-spider escapements.

Fig. 6.35 Star-wheel and worm escapements.

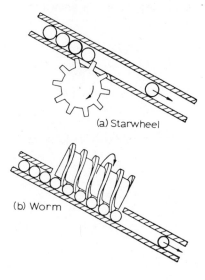

(a) Starwheel

(b) Worm

Drum escapements may be driven continuously or indexed, but usually an indexing mechanism is preferable because difficulties may be encountered in attempting to synchronize a continuous drive to meet the requirements of the workhead. Of the six types of drum escapement described (Figs. 6.33 through 6.35), the first four are effectively rotary slide escapements and the latter two, rotary ratchet escapements.

6.3.4 Gate Escapements

The gate escapement is seldom used as a means of regulating the flow of parts to an automatic workhead on a mechanized assembly machine. Its main use is in providing an alternative path for parts, and in this capacity it is often used for removing faulty parts from the main flow. However, one type of gate escapement, shown in Fig. 6.36, can be used to advantage on certain types of part when it is necessary to provide two equal outputs from a single feed track input. It is clear from the figure that although this device is usually referred to as an escapement, it does not regulate the flow of parts and further escapements would be necessary on the delivery tracks to carry out this function.

6.3.5 Jaw Escapements

Jaw escapements are particularly useful in automatic assembly applications where some forming process is necessary on the part after it has been placed in position in the assembly. Figure 6.37a shows an

Fig. 6.36 Gate escapement.

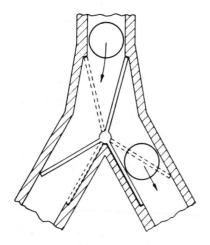

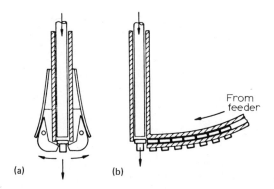

From
feeder

(a)   (b)

Fig. 6.37 Jaw escapement (a) and assisted gravity feed part placing
mechanism (b).

example of this type. The part is held by the jaws until the actuator,
in this case a punch, forces the part through the jaws. The punch
then performs the punching operation, returns, and allows another
part into the jaws. The device can, of course, be used purely as an
escapement.

6.4 Parts-Placing Mechanisms

Two simple types of parts-placing mechanisms have already been described and illustrated in Figs. 6.27 and 6.28. In the first two examples, the parts are taken from the feed track by the work carrier,
and in the other cases the parts are fed by gravity into pockets on
a rotary index table. These special applications, however, can only
be used for a very limited range of parts and by far the most widely
used parts-placing mechanism is a conventional gravity feed track
working in conjunction with an escapement. This system of parts
placing is probably the cheapest available but has certain limitations.
First, it may not be possible to place and fasten parts at the same
position on the machine, because of interference between the feed
track and the workhead. This would necessitate a separate workstation
for positioning the part, which would result in an increase in the length
of the machine. It then becomes necessary to retain the part in its
correct orientation in the assembly during transfer. Second, if a
close fit is required between the part and the assembly, the force
due to gravity may not be sufficient to ensure that the part seats
properly. Third, if the part cannot be suitably chamfered, the gravity
feed track may not give the precise location required. However, for
placing of screws and rivets prior to fastening, which together form

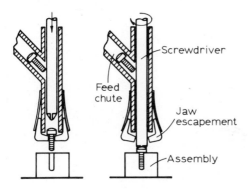

Fig. 6.38 Parts placing mechanism for automatic screwdriver.

a large proportion of all parts-placing requirements, and where the tool activates the escapement and applies the required force to make assembly possible, the gravity feed track is invariably used. An example of automatic screw placing and driving is shown in Fig. 6.38.

A further example of the assisted gravity feed type is shown in Fig. 6.37b, where a part is being positioned in the assembly. With this device, the feed track positions the part vertically above the assembly. The part is then guided into position in the assembly by a reciprocating guide rod. With these systems, it is common to fit the escapement at the point where the guide rod operates and to use the guide rod as the escapement activator. In some applications, the part may be positioned above the assembly by means of a slide escapement and then guided into position using a guide rod. This system is commonly referred to as the *push and guide* system of parts placing.

For situations where the placing mechanism has to be displaced from the workstation location, the *pick and place* system is often used. The basic action of this system is shown in Figs. 6.39 and 6.40, where it can be seen that the part is picked up from the feed track by means of a mechanical, magnetic, or vacuum hand, depending on the particular application, placed in position in the assembly, and then released. The transfer arm then returns along the same path to its initial position. Proprietary pick and place mechanisms operate in a variety of ways. Some pick vertically, transfer along a straight path horizontally, and place vertically, as shown in Fig. 6.39a. Others pick vertically, transfer around the arc of a circle in a horizontal plane, and place vertically, as shown in Fig. 6.39b; a third type picks and places by rotary motion of the transfer arm in a vertical plane and throughout transfer, the hand remains vertical (Fig. 6.40a). Finally, a variation of this latter type is where the pickup head is fixed to the arm and the part is turned

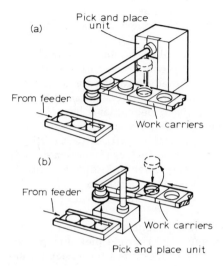

Fig. 6.39 Pick and place units that lift and position vertically.

Fig. 6.40 Pick and place units that move the part along the arc of a circle.

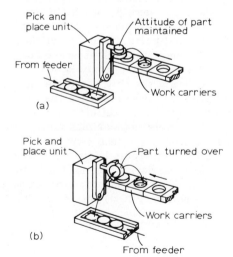

over during the operation (Fig. 6.40b). When operating correctly, the first two examples can mate parts that have no chamfers and close assembly tolerances, whereas the latter two examples, although the simplest systems, will operate only when the length of the vertical contact between the mating parts is small.

On high-speed operations the pick and place mechanisms are usually mechanically actuated and are mechanically connected to the entire assembly operation. Each system is designed and built for a particular application. However, because of the similarity between many of the assembly operations, modular pick and place mechanisms have been developed. These systems are basically off-the-shelf items which can handle a large percentage of assembly operations with little modification. These systems have been developed in-house in larger companies, are part of the basic machine for some assembly machine builders, and are available as separate items from individual manufacturers. The modular concept has been successful because many assembly operations consist of the same sequence of simple motions.

Recent studies have indicated that most subassemblies and some complete assemblies can be put together with a single pick and place unit, with 3 or 4 degrees of freedom, as represented by a rectangular or cylindrical coordinate system with a twisting motion about the major axis. Such a mechanism needs an elaborate control system to handle the larger variety of sequential motions. These systems are called *robots* and are a combination of a computer control system and a mechanical arm. Although these systems are too complex and expensive for mass production, it is believed that this unit will open a new era for the development of the reprogrammable assembly station. This station will consist of a reprogrammable pick and place unit or robot surrounded by families of correctly oriented parts. The idea of the station stems from group technology. The station will assemble complete families of similar assemblies. The carburetor family is one example. The individual models in the family are only variations of the main family member. For carburetors the base units are the same but the tubing is connected differently to make the different models.

A simpler, less expensive yet programmable modular pick and place unit uses pneumatic cylinders, positive stops, and limit switches and the assembly accuracy is developed using guide pins in the hand which slide in hardened guide holes in the work carriers. The motion of the unit can be controlled by pneumatic digital logic elements, programmable controllers, programmable read-only memories (PROM), or microcomputers.

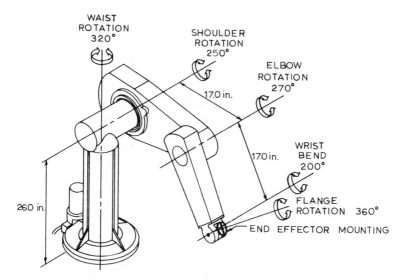

WAIST
ROTATION
320°

SHOULDER
ROTATION
250°

ELBOW
ROTATION
270°

17.0 in.

WRIST
BEND
200°

17.0 in.

FLANGE
ROTATION 360°

26.0 in.

END EFFECTOR MOUNTING

Fig. 6.41 PUMA 500 Series robot. (Courtesy of Unimation Inc.)

6.5 Assembly Robots

Considerable attention is currently being given to applying automatic assembly techniques developed for mass production to the area of batch production, where most products are made. One concept generally thought applicable is to replace the dedicated workstation with a programmable workstation, and much effort and study has been devoted to substituting a robot for a conventional pick and place unit. This robot is typically a single mechanical arm with its motion under computer control.

Several robots, which are currently available, can be used for assembly in this way. One such robot, called the PUMA* (programmable universal manipulator for assembly), has emerged from the cooperative efforts of General Motors Corporation and Unimation, Inc., and is shown in Fig. 6.41. As the name implies, this robot has been designed for assembly. It is smaller and lighter than most industrial robots and occupies a volume roughly equivalent to a human operator. It can handle parts with a mass of up to 2.5 kg, has a location repeatability of ±0.1 mm, and can apply a static assembly force of 60 N (Newton: the basic unit of force in SI units, equal to 0.22 pounds force).

---

*P. F. Rogers, "The PUMA, the VAL Language, and Programmable Assembly," Conference of the Swiss Federation of Automatic Control on Industrial Robots, Mar. 1980.

One of the advantages of the PUMA is its ease of programming. Most industrial robots and robot applications fail to use the robot's reprogramming capabilities. They are often programmed for a special task once in their lifetime. However, the applications for the PUMA are envisioned to be entirely different. As the economic analysis in Chapter 7 indicates, future applications of programmable assembly may cover a wide range of products, and a single-station system should be readily adaptable to design changes and product style variations. For this work, reprogramming is a necessity, but conventional robot programming requires such a significant time that it could adversely affect robot economics.

One way of reducing programming time uses the concept of hierarchical control. Such a system has been proposed by the National Bureau of Standards* and is represented schematically in Fig. 6.42. With this system there are five levels of control, with each upper level showing increased sophistication in computer software but reduced effort in task programming.

Level 1 represents the basic servo control of a conventional robot. The inputs to this level of control are the desired joint positions, the current joint positions, and velocity feedback. The outputs are the joint errors that form the drive signals to the servo actuators.

The output from level 2, the primitive function control level, is the input to level 1. This control level generates trajectory information, handles sensory feedback, and allows the programmer to work with a real-world coordinator system, which is more applicable to the problem and easier to handle than the joint coordinate system of the robot. Typical programming commands to this level would be APPROACH, DETECT, or GRASP.

Level 3 in the hierarchy allows several primitive commands to be combined into a single command. A command such as GET PART A would be equivalent to the primitive commands APPROACH A, DETECT, and GRASP. Similarly, commands to level 4 would be equivalent to a series of commands to level 3. This fourth level, called workstation control, could respond to a command such as ASSEMBLE A, TEST B, and so on. The top level, level 5, is designed to work with system commands such as MAKE K1.

As an example, suppose that this last command, MAKE K1, is issued to a single assembly robot which is surrounded by magazines of parts and is capable of assembling several models like model K1. The system control (level 5) sends a series of commands to the workstation control (level 4) to assemble and test for the individual components. One of these statements, ASSEMBLE A, is sent to the workstation control,

---

*A. J. Barbera, National Bureau of Standards Special Publication 500-23, Dec. 1977.

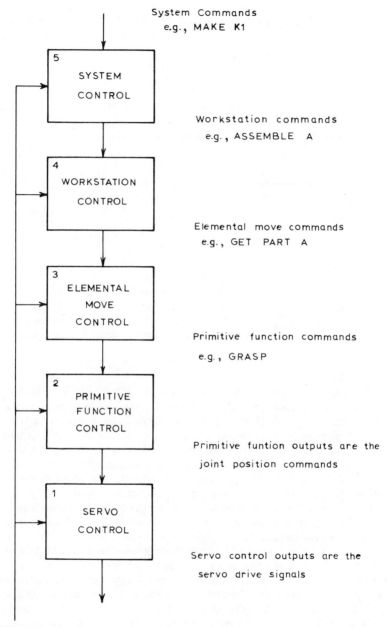

System Commands
e.g., MAKE K1

Workstation commands
e.g., ASSEMBLE A

Elemental move commands
e.g., GET PART A

Primitive function commands
e.g., GRASP

Primitive funtion outputs are the
joint position commands

Servo control outputs are the
servo drive signals

FEEDBACK

Fig. 6.42 Five-level robot control hierarchy. (Adapted from A. J. Barbera, National Bureau of Standards Special Publication 500-23, Dec. 1977.)

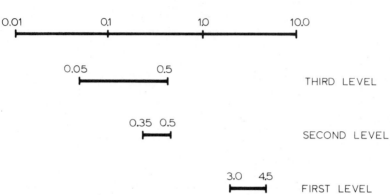

Fig. 6.43 Results of programming an insertion task. (Adapted from A. J. Barbera, National Bureau of Standards Special Publication 500-23, Dec. 1977.)

which initiates a series of commands such as GET PART A, which is sent to the elemental move control (level 3), which, in turn, issues the commands APPROACH A, DETEC A, GRASP, and so on, to the primitive function control (level 2). The location of part A has been stored in the computer at the time of the initial setup. Using this information, the computer can generate the joint position commands to the servo control (level 1), which can activate the robot motions.

Although this five-level robot control hierarchy is not available at present, the concept promises to reduce programming time significantly. Experiments conducted at the National Bureau of Standards show that programming a robot with three control levels of the hierarchy results in a reduction in programming time as great as two orders of magnitude.* The results of programming an insertion task are shown in Fig. 6.43. This insertion was vertically downward and the robot, like the PUMA, must coordinate several joint motions simultaneously to produce this straight-line motion.

Programming at level 1 required moving the robot through a number of successive points along the straight-line path using the individual servos, which took as long as 4.5 min. With level 2, the programmer used a joystick to produce motions in an XYZ coordinate system.

---

*Ibid.

VAL Pick and Place Program

```
PROGRAM PICK
    APPRO A, 100
    MOVES A
    CLOSEI
    DEPART 100
    APPRO B, 100
    MOVES B
    OPENI
    DEPART 100
    RETURN
```

VAL Palletizing Subroutine

```
PROGRAM PALLET
        SET HOLE = CORNER
        SETI COL = 1
100     SETI ROW = 1
200     GOSUB PICK
        SHIFT HOLE BY 50
        SETI ROW = ROW + 1
        IF ROW LT 4 THEN 200
        SHIFT HOLE BY -150, 50
        SETI COL = COL + 1
        IF COL LT 5 THEN 100
```

Fig. 6.44 VAL assembly programs. (Adapted from P. F. Rogers, "The PUMA, the VAL Language, and Programmable Assembly," Conference of the Swiss Federation of Automatic Control on Industrial Robots, Mar. 1980.)

Again, it was necessary to teach the motion by moving the robot through a number of points along a vertical path, but the joystick control made vertical motion easy and reduced the programming time to 0.5 min. Alternatively, a computer program could be written as a series of elemental move commands. These commands would be, for example, APPRO A, 100 and MOVES A. The first command has the robot move to a position 100 mm above point A, while the second command causes the robot to move in a straight path to point A.

Obviously, the computer must know the location of point A, which represents the end point of the insertion. Programming at level 3 could be simply INSERT A, 100 and it is not surprising that programming time at this level is as low as 0.05 min.

The PUMA robot uses concepts associated with the first three levels in the hierarchy and utilizes a programming language called VAL with an LSI-11 controller. Two examples of programs written in VAL are shown in Fig. 6.44. The first program is for the pick and place motions where a part is picked up at location A and placed at point B following three sides of a rectangular path. Again, the location of points A and B must be preset in computer memory.

The second program is for removing parts from a 3 × 2 pallet. It follows a format similar to FORTRAN and utilizes the previous program as a subroutine. The adjacent parts on the pallet are 50 mm from one another. This program can accommodate a variety of pallet sizes and spacing by simple editing, which is another advantage of this hierarchical approach.

Alternatively, miscellaneous accessories are being studied and/or developed to increase robot assembly performance. The Remote Center Compliance (RCC) allows insertion with small tolerances which would otherwise require special fixtures.* The use of force, light, and tactile sensors are being explored to give the robot information for simple decision, and binary vision is being considered as a solution to the parts feeding and orienting problem. Interestingly, little effort has been devoted to gripper design, even though the economic analysis in Chapter 7 indicates that this is an area of great potential. Design of parts and products for assembly, on the other hand, is almost universally accepted as the area of greatest promise if programmable assembly is to make a significant contribution to productivity in the near future.

Finally, it is important to realize that this technology can have sweeping social effects, and perhaps the most important consideration for the widespread acceptance of robots in the workplace is education— education to dispel the myths and promulgate the truths about robots and man's future.

---

*D. E. Whitney and J. L. Nevins, "What Is the Remote Center Compliance (RCC) and What Can It Do?," Proceedings of 9th International Symposium on Industrial Robots, Washington, D.C., 1979.

Chapter 7

Performance and Economics of Assembly Systems

Multistation automatic assembly machines may be classified into two main groups according to the system used to transfer assemblies from workstation to workstation. The larger of the two groups includes those assembly machines that transfer all the work carriers simultaneously. These are known as indexing machines, and a stoppage of any individual workhead causes the whole machine to stop. In the other group of machines, which are known as free-transfer assembly machines, the workheads are separated by buffer stocks of assemblies and transfer to and from these buffer stocks occurs when the particular workhead has completed its cycle of operations. Thus, with a free-transfer machine, a fault or stoppage of a workhead will not necessarily prevent another workhead from operating because a limited supply of assemblies will usually be available in the adjacent buffer stocks.

One of the principal problems in applying automation to the assembly process is the loss in production, resulting from stoppages of automatic workheads when defective component parts are fed to the machine. With manual workstations on an assembly line, the operators are able to discard defective parts quickly and little loss of production occurs. However, a defective part fed to an automatic workhead can, on an indexing machine, cause a stoppage of the whole machine, and production will cease until the fault is cleared. The resulting downtime can be very high with assembly machines having several automatic workheads. This can result in a serious loss in production and a consequent increase in the cost of assembly. The quality levels of the parts to be used in automatic assembly must therefore be considered when an assembly machine is designed.

In the following sections, a study is made of the effects of the quality levels of parts on the performance and economics of assembly machines.

7.1 Indexing Machines

7.1.1 Effect of Parts Quality on Downtime

In the following analysis, it will be assumed that an indexing machine having n automatic workheads and operating on a cycle time of t seconds is fed with parts having, on average, a ratio of defective parts to acceptable parts of x. It will also be assumed that a proportion m of the defective parts will cause machine stoppages and further, that it will take an operator T seconds, on average, to locate the failure, remove the defective part, and restart the machine.

With these assumptions, the total downtime due to stoppages in producing N assemblies will be given by mNxnT. Each time the machine indexes, all assembly tasks are completed and one assembly is delivered from the machine, hence, the machine time to assemble N assemblies is Nt seconds, and thus the proportion of downtime D on the machine is given by

$$D = \frac{downtime}{assembly\ time + downtime}$$

$$= \frac{mxnNT}{Nt + mxnNT} = \frac{mxn}{mxn + t/T} \tag{7.1}$$

In practice, a reasonable value of the machine cycle time t would be 6 s, and experience shows that a typical value for the average time taken to clear a fault is 30 s. With these figures, the ratio t/T will be 0.2 and Fig. 7.1 shows the effect of variations in the mean quality level of the parts on the downtime for indexing machines with 5, 10, 15, and 20 automatic workheads. (It is assumed in this example that all defective parts will produce a stoppage of the machine and thus m = 1.)

For standard fasteners such as screws, which are often employed in assembly processes, an average value for x would be between 0.01 and 0.02. In other words, for every 100 acceptable screws, there would be between one and two defective ones. A higher quality level is generally available, but with screws, for example, a reduction of x to 0.005 may double their price and seriously affect the cost of the final assembly. It will be seen later that a typical economic value for x is 0.01, and Fig. 7.1 shows that with this value the downtime on an assembly machine having 10 automatic workheads is 0.33 of the total time available. These results show why it is rarely economical to use indexing machines having a large number of automatic workheads. They also illustrate why, in practice, it is common to allow for a total downtime of 0.5 (50%) when considering the use of an indexing assembly machine.

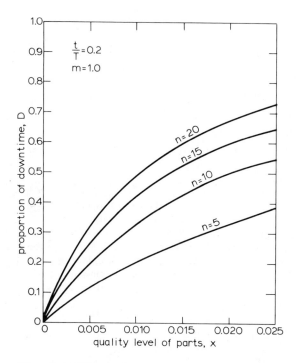

Fig. 7.1 Effect of parts quality on indexing machine downtime. n is the number of automatic workheads, t the machine cycle time, T the time to correct machine fault, and m the proportion of defective parts causing machine stoppage.

7.1.2 Effects of Parts Quality on Production Time

In the example above it was assumed that all the defective parts fed to the automatic workheads would stop the machine [that is, $m = 1$ in Eq. (7.1)]. In practice, however, some of these defective parts would pass through the feeding devices and automatic workheads but would not be assembled correctly and would result in the production of an unacceptable assembly. In this case the effect of the defective part would be to cause downtime on the machine equal to only one machine cycle. The time taken to produce N assemblies, whether these are acceptable or not, is given by (Nt + mNxnT), and if $m < 1$, only about $(N - (1 - m) \, xnN)$ of the assemblies produced will be acceptable. The average production time t_{pr} of acceptable assemblies is therefore given by

$$t_{pr} = \frac{Nt + mNxnT}{N - (1 - m)xnN}$$

$$= \frac{t + mxnT}{1 - (1 - m)xn} \tag{7.2}$$

Taking typical values of $x = 0.01$, $t = 6$ s, $T = 30$ s, and $n = 10$, Eq. (7.2) becomes

$$t_{pr} = \frac{30(2 + m)}{9 + m} \tag{7.3}$$

Equation (7.3) is plotted in Fig. 7.2 to show the effect of m on t_{pr}, and it can be seen that for a maximum production rate of acceptable assemblies, m should be as small as possible. In other words, when designing the workheads for an indexing assembly machine when a high production rate is required, it is preferable to allow a defective part to pass through the feeder and workhead and "spoil" the assembly rather than allow it to stop the machine. However, in practical circumstances, the cost of dealing with the unacceptable assemblies produced by the machine must be taken into account, and this will be considered later.

For the case where the defective parts always stop the machine, m is equal to 1, and Eq. (7.2) becomes

$$t_{pr} = t + xnT \tag{7.4}$$

Fig. 7.2 Variation in production time t_{pr} with changes in the proportion of defective parts causing a machine fault. n is the number of automatic workheads, x the ratio of defective to acceptable parts, t the machine cycle time, and T the time to correct machine fault.

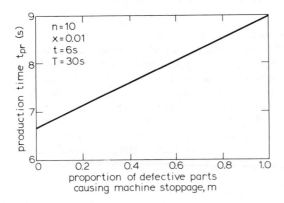

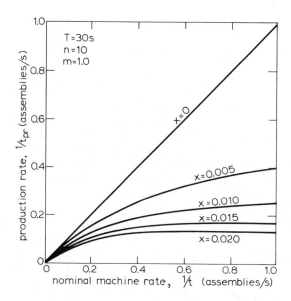

Fig. 7. 3 Effect of parts quality level on indexing machine production rate. n is the number of automatic workheads, T the time to correct machine fault, and m the proportion of defective parts causing machine stoppage.

Figure 7.3 now shows how the production rate $(1/t_{pr})$ is affected by changes in nominal machine rate $(1/t)$ for various values of x and for typical values of T = 30 s and n = 10. It can be seen that when x is small, the production rate approaches the machine speed. However, in general, x will lie within the range 0.005 to 0.02, and under these circumstances it can be seen that for high machine speeds (short cycle times) the production rate tends to become constant. Alternatively, it may be stated that as the cycle time is reduced for otherwise constant conditions, the proportion of downtime increases and this results in a relatively small increase in the production rate. This explains why it is rarely practicable to have indexing assembly machines working at very high speeds.

7.1.3 Effect of Parts Quality on the Cost of Assembly

The total cost C_t of each acceptable assembly produced on an assembly machine (where each workhead assembles one part) is given by the sum of the costs of the individual parts $C_1 + C_2 + C_3 + \cdots + C_n$ plus the cost of operating the machine for the average time taken to produce one acceptable assembly. Thus,

$$C_t = M_t t_{pr} + C_1 + C_2 + C_3 + \cdots C_n \qquad (7.5)$$

where M_t is the total cost of operating the machine per unit time and includes operators' wages, overhead charges, actual operating costs, machine depreciation and the cost of dealing with the unacceptable assemblies produced, and t_{pr} is the average production time of acceptable assemblies and may be obtained from Eq. (7.2).

In estimating M_t it will be assumed that a machine stoppage caused by a defective part will be cleared by one of the operators employed on the machine and that no extra cost will be entailed other than that due to machine downtime. Further, it will be assumed that if a defective part passes through the workhead and spoils an assembly, it will take an extra operator T_c seconds to dismantle the assembly and replace the nondefective parts back in the appropriate feeding devices.

Thus, the total operating cost M_t is given by

$$M_t = M + P_u T_c W \qquad (7.6)$$

where M is the cost of operating the machine per unit time if only acceptable assemblies are produced and W is the operator's rate, including overhead. The number of unacceptable assemblies produced per unit time is denoted by P_u and is given by

$$P_u = \frac{(1 - m)xn}{t + mxnT} \qquad (7.7)$$

Substitution of Eq. (7.7) in Eq. (7.6) gives

$$M_t = M + \frac{(1 - m)xnT_c W}{t + mxnT} \qquad (7.8)$$

In estimating the cost C_i of an individual component part, it will be assumed that this can be broken down into (1) the basic cost of the part irrespective of quality level, and (2) a cost that is inversely proportional to x and which will therefore increase for better-quality parts. Thus, the cost of each part may be expressed as

$$C_i = A_i + \frac{B}{x} \qquad (7.9)$$

In this equation B is a measure of the cost due to quality level and for the purposes of the present analysis will be assumed constant regardless of the basic cost A_i of the parts.

If Eqs. (7.2), (7.8), and (7.9) are now substituted into Eq. (7.5), the total cost C_t of each acceptable assembly becomes, after rearrangement,

$$C_t = \frac{M(t + mxnT) + (1 - m)xnT_cW}{1 - (1 - m)xn} + \sum_{i=1}^{n} A_i + \frac{nB}{x} \qquad (7.10)$$

Equation (7.10) shows that the total cost of an assembly can be broken down as follows:

1. A cost that will decrease as x is reduced
2. A cost that is constant
3. A cost that will increase as x is reduced

It follows that for a given situation, an optimum value of x will exist that will give a minimum cost of assembly. For the moment, the optimum value of x will be considered for the case where m = 1 (that is, where all defective parts cause a stoppage of the machine).
  With m = 1, Eq. (7.10) becomes

$$C_t = Mt \quad + \quad MxnT \quad + \quad \frac{nB}{x} \quad + \quad \sum_{i=1}^{n} A_i \qquad (7.11)$$

cost of assembly operations	cost of downtime	cost of parts quality	basic cost of parts

Equation (7.11) is now differentiated with respect to x and set equal to zero, yielding the following expression for the optimum value of x giving the minimum cost of assembly:

$$x_{opt} = (\frac{B}{MT})^{1/2} \qquad (7.12)$$

It is interesting to note that for a given assembly machine, where M and B are constants, the optimum quality level of the parts used is dependent only on the time taken to clear a defective part from a workhead.
  Figure 7.4 shows how the cost of a part might increase as the quality level is improved. In this case, A_i = \$0.002 and B = \$0.00003. If typical values of M = 0.01 \$/s (36 \$/h) and T = 30 s are now substituted in Eq. (7.12), the corresponding optimum value of x is 0.01 (1%). Equation (7.12) may now be substituted in Eq. (7.11) to give an expression for the minimum cost of assembly:

$$C_{t(min)} = Mt + 2n(MBT)^{1/2} + \sum_{i=1}^{n} A_i \qquad (7.13)$$

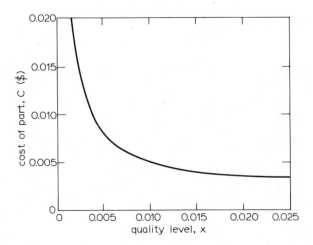

Fig. 7.4 Typical relationship between part quality and cost.

With $t = 6$ s and $n = 10$, the cost of assembling each complete set of parts [the first two terms in Eq. (7.13)] would be $0.12. Half of this cost would be attributable to the assembly operation itself and the other half attributable to the increased cost due to parts quality and the corresponding cost of machine downtime. Figure 7.5, where the first three terms in Eq. (7.11) are plotted, shows how these individual costs would vary as the quality level x varies, using the numerical values quoted in the example above. In this case, if parts having 0.02 defective were to be used instead of the optimum value of 0.01, the cost of assembly would increase by approximately 12%. This is a variation of $0.015 per assembly and, with an average production time of 9 s [calculated from Eq. (7.2)], represents an extra expense of approximately $12,000 per year per shift.

In the analysis above, it was assumed that all defective parts would stop the machine. If, instead, it were possible to allow these parts to pass through the automatic devices and spoil the assemblies, the cost of assembly could be obtained by substituting $m = 0$ into Eq. (7.10). Thus,

$$C_t = \frac{Mt + xnT_cW}{1 - xn} + \sum_{i=1}^{n} A_i + \frac{nB}{x} \tag{7.14}$$

Again an optimum value of x arises and is found by differentiation of Eq. (7.14) to be

$$x_{opt} = \left[n + \left(\frac{T_c W}{B} + \frac{Mt}{B} \right)^{1/2} \right]^{-1} \qquad (7.15)$$

Taking $W = 0.002$ \$/s, $T_c = 60$ s, and the remaining figures as before, x_{opt} is found to be approximately 0.011. From Eq. (7.14) the minimum cost of assembly is \$0.109, which represents a savings of 9% on the cost of assembly compared with the preceding example.

It is possible to draw two main conclusions from the examples discussed. First, for the situation analyzed, it would be preferable to allow a defective part to pass through the workhead and spoil the assembly rather than to allow it to stop the indexing machine. This would not only increase the production rate of acceptable assemblies but would also reduce the cost of assembly. Second, an optimum quality level of parts always exists that will give minimum cost of assembly.

7.2 Free-Transfer Machines

A free-transfer machine always gives a higher production rate than the equivalent indexing machine. This is because the buffer storage of assemblies available between workheads will, for a limited time, allow the continued operation of the remaining workheads when one has stopped. Provided that the buffer storage is sufficiently large, the stopped workhead can be restarted before the other workheads are affected and the downtime on the machine approaches the downtime

Fig. 7.5 Effect of parts quality level on assembly costs.

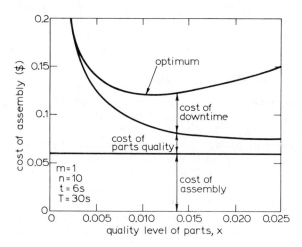

on the workhead that has the most stoppages. The following analysis
will show that, even with relatively small buffer storage, the production rate of the machine can be considerably higher than that obtained
with the equivalent indexing machine.

It will be assumed in the analysis that all the workheads on a free-
transfer machine are working at the same cycle time of t seconds.
Each workhead is fed with parts having the same quality level of x
(where x is the ratio of defective to acceptable parts), and between
each pair of workstations is a buffer storage that is large enough to
store b assemblies. Any workhead on a free-transfer assembly machine
will be forced to stop under three different circumstances.

1. If a defective part is fed to the workhead and prevents the completion of its cycle of operations. In this case it will be assumed that
   an interval of T seconds elapses before the fault is cleared and
   the workhead restarted.
2. If the adjacent workhead up the line has stopped and the supply
   of assemblies in the buffer storage between is exhausted.
3. If the adjacent workhead down the line has stopped and the buffer
   storage between is full.

If two adjacent workheads have a assemblies in the buffer storage
between, then a fault in the first workhead will prevent the second
from working after a time lag of at seconds. A fault in the second
workhead will prevent the first from working after a time lag of $(b - a)t$
seconds. The analysis is based on the fact that over a long period
the average downtime on each workhead must be the same. The assumption is made that no workhead will stop while another is stopped. The
errors resulting from this assumption will become large when the quality
level of the parts is poor (large x) and with a large number of automatic workheads (large n). However, specimen calculations show that
these errors are negligible with practical values of x and n and produce
an overestimate of the machine downtime.

7.2.1 Performance of Five-Station Machine

Considering the first workhead of a five-station machine (Fig. 7.6)
producing N assemblies, Nx stoppages will occur if m is unity. If each
fault takes T seconds to correct, the downtime on the first workhead
due to its own stoppages is given by NxT. This same average downtime will apply to the second workhead, but the first will only be
prevented from working for a period of $Nx[T - (b - a_1)t]$ seconds,
if a_1 is the average number of assemblies held in the buffer storage
following the first workhead. Similarly, stoppages of the third workhead will only prevent the first from working for a period of $Nx[T -
(2b - a_1 - a_2)t]$ seconds (a_2 is the average number of assemblies held

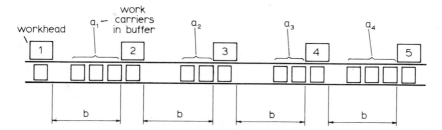

Fig. 7.6 Five-station free-transfer machine showing the average situation during operation for b = 4.

in the buffer storage following the second workhead). Similar expressions can be derived for the effects of the fourth and fifth workheads. Thus, it can be seen that the total downtime, d_1, on the first workhead while the machine produces N assemblies is given by

$$\frac{d_1}{Nx} = T + [T - (b - a_1)t] + [T - (2b - a_1 - a_2)t]$$

$$+ [T - (3b - a_1 - a_2 - a_3)t] + [T - (4b - a_1 - a_2 - a_3 - a_4)t]$$

$$(7.16)$$

It should be noted that if any term in square brackets is negative, it should be omitted.

Similar equations for the downtime on the remaining workheads may be derived and these are presented in Table 7.1.

If the condition is applied that the average downtime on all stations must be the same, then

$$d_1 = d_2 = d_3 = d_4 = d_5 \qquad (7.17)$$

and the five equations in Table 7.1 may be solved simultaneously to give the values of a_1 to a_4 inclusive in terms of b. In the example, the appropriate values are

$$a_1 = \frac{4b}{5}, \quad a_2 = \frac{3b}{5}, \quad a_3 = \frac{2b}{5}, \quad \text{and} \quad a_4 = \frac{b}{5} \qquad (7.18)$$

Substituting these values in the equations for the downtime d on each workhead gives

Table 7.1 General Equations for the Downtime d at Each Station of a Five-Station Assembly Machine

Workhead	Downtime due to first workhead	Second workhead	Third workhead	Fourth workhead	Fifth workhead
First: $\dfrac{d_1}{Nx}$	$= T$	$+ [T-(b-a_1)t]$	$+ [T-(2b-a_1-a_2)t]$	$+ [T-(3b-a_1-a_2-a_3)t]$	$+ [T-(4b-a_1-a_2-a_3-a_4)t]$
Second: $\dfrac{d_2}{Nx}$	$= [T-a_1t]$	$+ T$	$+ [T-(b-a_2)t]$	$+ [T-(2b-a_2-a_3)t]$	$+ [T-(3b-a_2-a_3-a_4)t]$
Third: $\dfrac{d_3}{Nx}$	$= [T-(a_1+a_2)t]$	$+ [T-a_2t]$	$+ T$	$+ [T-(b-a_3)t]$	$+ [T-(2b-a_3-a_4)t]$
Fourth: $\dfrac{d_4}{Nx}$	$= [T-(a_1+a_2+a_3)t]$	$+ [T-(a_2+a_3)t]$	$+ [T-a_3t]$	$+ T$	$+ [T-(b-a_4)t]$
Fifth: $\dfrac{d_5}{Nx}$	$= [T-(a_1+a_2+a_3+a_4)t]$	$+ [T-(a_2+a_3+a_4)t]$	$+ [T-(a_3+a_4)t]$	$+ [T-a_4t]$	$+ T$

$$\frac{d_1}{Nx} = T + [T - \frac{bt}{5}] + [T - \frac{3bt}{5}] + [T - \frac{6bt}{5}] + [T - \frac{10bt}{5}]$$

$$\frac{d_2}{Nx} = [T - \frac{4bt}{5}] + T + [T - \frac{2bt}{5}] + [T - \frac{5bt}{5}] + [T - \frac{9bt}{5}]$$

$$\frac{d_3}{Nx} = [T - \frac{7bt}{5}] + [T - \frac{3bt}{5}] + T + [T - \frac{3bt}{5}] + [T - \frac{7bt}{5}] \qquad (7.19)$$

$$\frac{d_4}{Nx} = [T - \frac{9bt}{5}] + [T - \frac{5bt}{5}] + [T - \frac{2bt}{5}] + T + [T - \frac{4bt}{5}]$$

$$\frac{d_5}{Nx} = [T - \frac{10bt}{5}] + [T - \frac{6bt}{5}] + [T - \frac{3bt}{5}] + [T - \frac{bt}{5}] + T$$

It is necessary at this stage to know the relative values of T and t. If, for example, the ratio T/t is 5, then the solution above applies only for $b \leqslant 2$; otherwise, some of the bracketed terms in Eq. (7.19) will become negative and must be omitted from the original equations in Table 7.1. For $b \leqslant 2$, the downtime d for the machine is now given from any Eq. (7.19). Thus,

$$\frac{d}{tNx} = 25 - 4b \qquad (7.20$$

The proportion of downtime D on the machine may now be obtained as follows:

$$D = \frac{\text{downtime}}{\text{assembly time} + \text{downtime}} = \frac{d}{Nt + d} \qquad (7.21)$$

and substitution of d from Eq. (7.20) gives

$$D = \frac{(25 - 4b)x}{1 + (25 - 4b)x} \qquad (7.22)$$

For $b \geqslant 3$ and with T/t = 5 some of the bracketed terms in Eq. (7.19) will become zero or negative and the corresponding terms in the equations in Table 7.1 must therefore be omitted. In this case, new values of a_1 to a_4 will be obtained from the new simultaneous equations and the magnitudes of the bracketed terms in the revised Eq. (7.19) must be reexamined. The process is repeated until no negative terms appear in the brackets in the revised form of Eq. (7.19).

Table 7.2 shows, for the ratio T/t = 5 and for increasing values of b, which terms in the original equations become negative and gives the corresponding values of a_1 to a_4.

Table 7.2 Effect of Buffer Stock Size b on Average Number of Assemblies a in Each Buffer Stock

Values of b	Omit terms in equations[a]					Values of a			
	d_1	d_2	d_3	d_4	d_5	$\dfrac{a_1}{b}$	$\dfrac{a_2}{b}$	$\dfrac{a_3}{b}$	$\dfrac{a_4}{b}$
0, 1, 2	—	—	—	—	—	4/5	3/5	2/5	1/5
3	5	5	—	1	1	3/4	7/12	5/12	1/4
4	4, 5	5	1, 5	1	1, 2	3/4	1/2	1/2	1/4
5, 6	4, 5	4, 5	1, 5	1, 2	1, 2	2/3	5/9	4/9	1/3
7	3, 4, 5	4, 5	1, 5	1, 2	1, 2, 3	22/35	19/35	16/35	13/35
8	3, 4, 5	4, 5	1, 5	1, 2	1, 2, 3	23/40	21/40	19/40	17/40
9	3, 4, 5	4, 5	1, 5	1, 2	1, 2, 3	24/45	23/45	22/45	21/45
≥10	2, 3, 4, 5	1, 3, 4, 5	1, 2, 4, 5	1, 2, 3, 5	1, 2, 3, 4	1/2	1/2	1/2	1/2

[a]See Table 7.1.

Table 7.3 Relationship Between Buffer Stock Size b and the Factor K

b	K
0	25.0
2	17.0
4	11.0
6	8.33
8	6.6
10	5.0

Having determined the values of a_1 to a_4 for each value of b, the percentage downtime on the machine may be calculated in the manner described above. In general for $0 \leqslant b \leqslant 10$, the machine downtime d may be expressed as

$$d = KNxt \qquad (7.23)$$

where K is a factor that depends on the values of T/t and b. Table 7.3 gives the values of K for various values of b when $T/t = 5$.
The proportion of downtime D is now given by

$$D = \frac{d}{Nt + d} = \frac{Kx}{1 + Kx} \qquad (7.24)$$

For $b \geqslant 10$ all bracketed terms are omitted from the equations in Table 7.1 and the equation for the downtime on each workhead becomes

$$\frac{d}{Nx} = T \qquad (7.25)$$

The downtime on the machine equals the downtime on an individual workhead, and thus for $b \geqslant 10$

$$D = \frac{5x}{1 + 5x} \qquad (7.26)$$

Figure 7.7 shows how the proportion of downtime is affected by the size of the buffer stocks for values of x = 0.005, 0.01, and 0.02. It can be seen that significant improvements in performance can be obtained with only small buffer stocks and that with large buffer stocks the machine downtime proportion approaches one-fifth of the downtime proportion on an equivalent five-station indexing machine (given by b = 0).

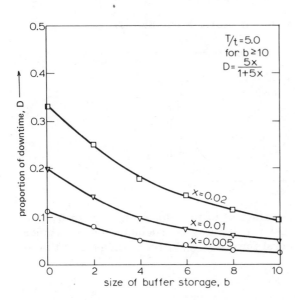

Fig. 7.7 Effect of buffer storage size on proportion of downtime for a give-station free-transfer machine.

In theory, when the size of the buffer storage b is greater than or equal to $2T/t$, the workheads are completely isolated one from the other and further increases in the size of b are futile. However, the greatest benefit occurs with the smaller buffers. So as a practical guide, it is often assumed that for a good free-transfer assembly machine design, b equals T/t.

7.2.2 Performance of Free-Transfer Machines When the Quality Levels of the Parts Vary from Station to Station

For simplicity in the foregoing analysis, it was assumed that the quality level of the parts was identical for each workhead (that is, $x_1 = x_2 = x_3 = x_4 = x_5$). In a practical situation, this will not generally be the case, but the procedure described above may still be employed to predict machine performance. It is clear, however, that unlike the indexing machine, the downtime on a free-transfer machine may be affected by the relative positions of the workheads when the quality levels of the parts vary from workhead to workhead. It can be argued that, in this case, the workheads with the worst quality levels (large x) should be placed as near to either end of the machine as possible where they will influence a minimum number of the remaining workheads.

This is now illustrated by a simple example. It will be assumed that a three station in-line free-transfer assembly machine has buffer storage between stations capable of holding four assemblies each. The parts fed to the workheads have quality levels x_1, x_2, and x_3 of 0.01, 0.02, and 0.01 defective, respectively. Assuming that $T/t = 5$, the downtime for this arrangement will be compared with that obtained when the workhead dealing with parts having 0.02 defective is placed at one end of the machine. The appropriate equations for the downtime at each workhead are

$$\frac{d_1}{Nt} = x_1 5 + x_2[5 - (4 - a_1)] + x_3[5 - (8 - a_1 - a_2)]$$

$$\frac{d_2}{Nt} = x_1[5 - a_1] + x_2 5 + x_3[5 - (4 - a_2)] \qquad (7.27)$$

$$\frac{d_3}{Nt} = x_1[5 - (a_1 + a_2)] + x_2[5 - a_2] + x_3 5$$

1. Writing $x_1 = x_3 = 0.01$ and $x_2 = 0.02$ in Eq. (7.27) and since $d_1 = d_2 = d_3$, the values of a_1 and a_2 may be found. Thus,

    $a_1 = 3$ and $a_2 = 1$

    With these values none of the terms in brackets in Eq. (7.27) should be omitted, and thus from any Eq. (7.27), $d = 0.14Nt$. Hence, the downtime proportion D is given by

    $$D = \frac{0.14Nt}{0.14Nt + Nt} = 0.123$$

2. Writing $x_1 = x_2 = 0.01$ and $x_3 = 0.02$ in Eq. (7.27), the values of a_1 and a_2 become

    $a_1 = 3$ and $a_2 = 2$

    Again, none of the terms in Eq. (7.27) should be omitted, and thus it is found that $d = 0.13Nt$. Hence,

    $$D = \frac{0.13Nt}{0.13Nt + Nt} = 0.115$$

The results of this example indicate that workheads dealing with parts having poor quality should be situated as near the end of the machine as possible.

It should be pointed out that the method of analysis for free-transfer machines presented in this chapter is only approximate and assumes that the stoppages due to defective parts will not occur in quick succession on the same workhead. Experimental work on an analog of a free-transfer machine* has shown that the analysis gives a good approximation to the true performance of a machine when $b \leqslant T/t$.

It should also be noted that the analysis applies only to a machine where relatively large buffer storage of empty work carriers are held in the return conveyor, between the last and the first workheads. If this were not the case, the basic equations would need modification to allow for a direct influence of stoppages at the last station on the performance of the first, and vice versa.

Before proceeding to analyze the economics of assembly systems further, it will be necessary to explain the basis for the economic comparisons presented.

7.3 Basis of Economic Comparisons for Automation Equipment

An important constant Q used in the present analyses is defined as the equivalent cost of capital equipment used to replace one operator on one shift. Whether a company uses return on investment or payback period to estimate machine rates, the figure Q can be obtained. By using this figure, the cost comparisons made later are therefore valid, regardless of the method employed by a company in their economic analyses.

For example, the effective cost for a piece of capital equipment costing C is some factor f times C, where the factor f depends upon the accounting procedures in the company, the interest rate, inflation, and the company's general economic climate. If this piece of equipment is virtually equivalent to n operators working S shifts for y years at an effective annual wage rate (including overheads) of W, then

$$Cf = nSyW \qquad (7.28)$$

If Q is the amount that can be spent on capital equipment to replace one operator on one shift, then from Eq. (7.28),

$$Q = \frac{C}{nS} = \frac{yW}{f} \qquad (7.29)$$

Thus, the value of Q can be found by multiplying the annual wage rate by the number of years of truly useful life of the equipment and

*G. Hunter, unpublished work, University of Salford, Salford, England.

dividing by the factor f from above. Using Eq. (7.29), the annual cost per shift for this equipment is

$$nW = \frac{CW}{SQ} \qquad (7.30)$$

This parameter CW/SQ is an annual rate per shift for a piece of equipment that initially cost C but whose cost can be financed over a number of years. For many companies, the ratio SQ/W is relatively constant and is often in the range 1 to 3, with the higher values indicating situations where the introduction of automation is more easily justified.

If a simple piece of automation were being considered and it can do the job of one operator, the economic cost of the machine (including overhead) would be Q for one shift working, 2Q for two shifts working, and so on.

Taking a more complicated example, suppose that an operator is performing a series of assembly operations which involve n parts and the average assembly time per part is t_0 if no mechanical assistance is provided to the operator. If the operator is paid at the rate of W, the assembly cost per completed assembly will be

$$C_{pr} = Wnt_0 \qquad (7.31)$$

Suppose now that the operator is provided with oriented parts close at hand by means of automatic feeding devices (one for each part), each of which costs C_F and as a result, the average assembly time per part is reduced to t_0'. In this case, the cost of assembly per assembly will be

$$C_{pr} = Wnt_0' + \frac{nC_FWnt_0'}{SQ} \qquad (7.32)$$

An economic value of C_F (the cost of one feeder) can be obtained by equating Eqs. (7.31) and (7.32). Thus,

$$Wnt_0 = Wnt_0' \left(1 + \frac{nC_F}{SQ}\right)$$

or

$$C_F = \left(\frac{t_0}{t_0'} - 1\right)\frac{SQ}{n} \qquad (7.33)$$

If, in a particular situation, the ratio t_0/t_0' is equal to 2, this would mean that an operator's assembly time could be halved by providing oriented parts from automatic feeders. Also, supposing that two shifts are worked and Q is $20,000, then

$$C_F = \frac{2(20,000)}{n} \qquad (7.34)$$

where n is the number of parts in the assembly.

It can be seen that the smaller the number of parts, the more could be spent on each feeding device. For only two parts then, $20,000 could be spent on each of the two devices. This result is obvious because the extra output of assemblies could be produced by one extra operator each shift and the equivalent cost per operator is $20,000 (the value of Q). Another way of using Eq. (7.34) would be to take the known cost of a feeder (say $5000) and rearrange the equation to give the maximum number of parts per assembly for economic application of feeders.

Thus,

$$n = \frac{2(20,000)}{5000} = 8 \qquad (7.35)$$

Thus, with the figures employed, it would be worth considering automatic feeding of parts to the operator if the assembly contains no more than 8 parts.

In summary, whenever a cost per unit time is required for a piece of equipment, it can be obtained from the equation

$$M = \frac{CW}{SQ} \qquad (7.36)$$

where M is the equipment rate, C the capital cost of equipment (including overhead), W the assembly operator rate (including overhead), S the number of shifts, and Q the equivalent capital cost of one operator on one shift. Using this approach, the economics of free-transfer machines will be studied and compared with the economics of indexing machines.

7.4 Comparison of the Economics of Free-Transfer
        and Indexing Machines

When considering the economics of free-transfer machines, the cost of providing the buffer storage has to be taken into account. Ideally, it is necessary to develop a mathematical model where the size b of the buffer storage between each station is a variable and then to study how the assembly costs C_{pr} are affected by the magnitude of b. However, later analysis will show that when the number of parts to be assembled n is small, the free-transfer machine is uneconomic compared with an indexing machine. When n is large, the reverse is true. Thus, when n is small, the optimum value of b is zero (representing an indexing machine) and when n is large, the optimum value is close to that at which the various workheads do not affect one another significantly. Inspection of Fig. 7.7, which shows the effect of buffer stock size on the proportion of downtime, indicates that when $b \geqslant 10$, the proportion of downtime D on the machine becomes equal to the proportion of downtime on an individual workhead. In general, this is true if $b \geqslant 2T/t$, where T is the downtime due to one defective part and t is the workhead cycle time. However, analysis of the economics of individual machines shows that a value of b equal to T/t results in cost savings closely approaching those for $b = 2T/t$, and therefore to simplify analysis and to make rough cost comparisons, it will be assumed that, on a free-transfer machine, $b = T/t$ and that the resulting downtime for the machine approaches that for an individual station.

Hence, for a free-transfer machine of any size, the average production time can be estimated from

$$t_{pr} = t + xT \qquad (7.37)$$

where x is the parts quality level. It is, of course, assumed here that all defective parts will cause a stoppage of the workhead. Thus, Eq. (7.37) can be compared with Eq. (7.4) for an indexing machine.

In analyzing the economics of these machines, the following nomenclature will be used and the numerical values given in parentheses represent the estimates of the cost of the equipment used in the later comparisons:

C_T = cost of transfer device per workstation for an indexing machine (10 k\$)

C_B = cost of transfer device per space (workstation and buffer) for a free-transfer machine (5 k\$)

C_c = cost of work carrier (1 k\$)

C_F = cost of automatic feeding device and delivery track (5 k\$)

C_W = cost of workhead (10 k\$)

W_t = total rate for all operators engaged on the machine (0.006 \$/s representing one engineer and one assembly operator)

W = rate for one assembly operator (0.002 \$/s)

Q = equivalent cost of one assembly operator in terms of capital investment (30 k\$)

Although the numerical values assigned for equipment costs may seem high, it should be realized that these include the basic equipment costs, the assembly machine design, engineering and debugging costs, the cost of controls, and so on. It has been estimated* that basic equipment costs often form only 40% of their total costs.

7.4.1 Indexing Machine

For an indexing machine, the total rate M for the machine and operators will be given by

$$M = W_t + \frac{nW}{SQ}(C_T + C_W + C_F + C_c) \qquad (7.38)$$

The assembly costs per assembly are now given by

$$C_{pr} = Mt_{pr} \qquad (7.39)$$

Substituting for M from Eq. (7.38) and for t_{pr} from Eq. (7.4) into Eq. (7.39) yields

$$C_{pr} = (t + xnT)[W_t + \frac{nW}{SQ}(C_T + C_W + C_F + C_c)] \qquad (7.40)$$

7.4.2 Free-Transfer Machine

For a free-transfer machine, where the number of work carriers chosen is such that, on average, the machine will be half full, the total rate M will be given by

$$M = W_t + \frac{nW}{SQ}\left[C_W + C_F + (b + 1)(C_B + \frac{C_c}{2})\right] \qquad (7.41)$$

Substituting for b = T/t, t_{pr} from Eq. (7.37) and M from Eq. (7.41) into Eq. (7.39)

$$C_{pr} = (t + xT)\left[W_t + \frac{nW}{SQ}(C_W + C_F + (1 + \frac{T}{t})(C_B + \frac{C_c}{2})\right] \qquad (7.42)$$

Assuming a cycle time t of 6 s, a downtime T of 30 s, a quality level x of 0.01, and the numerical values of equipment and operator costs described earlier, Fig. 7.8 shows how the costs of production on the two machines vary with the size of the machine. It can clearly be seen that for small values of n (small machines), the indexing machine

*P. M. Lynch, "Economic Technological Modeling and Design Criteria for Programmable Assembly Machines," Report T-625, The Charles Stark Draper Laboratory, Inc., Cambridge, Mass., June 1976.

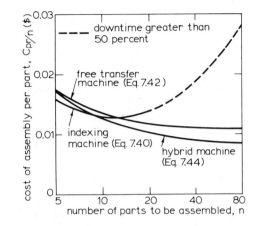

Fig. 7.8 Comparison of economics of multistation assembly machines.

is the more economical of the two, whereas for large values of n, the free-transfer machine is the more economical. The rapid rise in costs on the indexing machine is attributable entirely to the increasing downtime as the machine becomes larger. In fact, the region of the curve shown dashed represents downtime proportions of greater than 50%.

7.5 Multistation Hybrid System (Indexing Machines Linked by Buffer Storage)

For the assembly of products with large numbers of parts, a single indexing machine cannot be used because of the excessive downtime and a free-transfer machine tends to take up too much space because of buffer storage. A compromise between the two systems is a hybrid multistation machine whereby small indexing machines, usually of the rotary type, are interconnected by free-transfer conveyors.

    In such a system, if the buffer storage between indexing machines is adequate, the downtime for the system is equal to the downtime on one indexing unit. Hence, the production time is given by

$$t_{pr} = t + ixT \qquad (7.43)$$

where i is the number of stations on an individual indexing unit and t is the cycle time of the indexing units. Again assuming that the optimum buffer storage between units is T/t, the cost of assembly is

$$C_{pr} = t_{pr} \left[W_t + \frac{nW}{SQ} (C_T + C_F + C_c + C_W) + \frac{nW}{SQi} \frac{T}{t} (C_B + \frac{C_c}{2}) \right]$$

$$(7.44)$$

An optimum indexing unit size will arise giving minimum cost. This can be found by substitution of Eq. (7.43) into Eq. (7.44) and differentiation with respect to i. Thus,

$$i_{opt}^2 = \frac{(1/x)(C_B + C_c/2)}{(W_t SQ/nW) + C_T + C_F + C_c + C_W}$$

$$(7.45)$$

Using the same numerical values as before, Eq. (7.45) becomes

$$i_{opt}^2 = \frac{550}{(180/n) + 26}$$

$$(7.46)$$

As n varies from 10 to 80, the value of i varies only from 3.5 to 4.5. Clearly, in practice, the value of i must be an integer and for uniformity the ratio n/i should also be an integer. For the cost figures considered it would seem safe to assume four or five operations at each indexing unit. With i equal to 5, the effect of n on the cost of assembly given by Eq. (7.44) is plotted in Fig. 7.8 for comparison purposes. It can be seen that the multistation hybrid system is economic for most values of n. It should be remembered that, because of space requirements between units and because the output from a rotary indexing unit is adjacent to the input, a more complicated inter-unit transfer system might be required. The analysis does show, however, that during the design of a large assembly machine, the hybrid system should be considered.

The multistation assembly machines considered above, that is, indexing, free-transfer, and hybrid machines, are all special purpose or "dedicated" to one product. These are virtually the only types of machines to have been successfully applied in industry as yet. However, it should be realized that the output from one machine is on the order of 1 to 5 million assemblies per year and application of these systems is limited to mass production situations when the product design is stable for a few years. It should also be realized that mass production constitutes only about 15 to 20% of all production in the United States, and for this reason much attention is being given at present to the possibilities of automating batch assembly processes. It is clear that for these applications it will be necessary to employ programmable or flexible automation: in fact, assembly robots. Before analyzing the economics of such equipment, however, mathematical models for operator assembly schemes will be established for purposes of later comparison.

7.6 Multistation Operator Assembly

Here the assembly process is broken down into individual tasks and performed in sequence by a series of operators arranged in assembly line fashion. An individual operator will continually repeat the same operation or limited series of operations and the rate of output from the line is dependent on the time taken by the slowest operator.

One advantage of this scheme compared with a multistation assembly machine is that by providing each operator with more than one assembly task (several parts assembled at each station), the output from the assembly line can be matched more closely to the production rate required. Another advantage is that defective parts do not create the severe problems encountered on assembly machines. An operator can quickly recognize a defective part and discard it with little loss in production. If it is assumed that each defective part involves a simple repetition of the assembly task and that the average time taken by the slowest operator is t_0 seconds per task, the mean production time t_{pr} per assembly is given by

$$t_{pr} = kt_0(1 + x) \tag{7.47}$$

where k is the number of parts assembled by each operator.

Assuming that a free-transfer device with work carriers is provided and that each operator has two spaces (one buffer) on the transfer device to allow for minor delays in assembly time, the cost of assembly is given by

$$C_{pr} = \frac{nWt_{pr}}{k}\left(1 + \frac{2C_B + C_c}{SQ}\right) \tag{7.48}$$

Substituting Eq. (7.47) into Eq. (7.48), thereby eliminating k, and using the same numerical values for the cost terms as before and an assembly time per part t_0 of 10 s, the assembly costs per part C_{pr}/n become \$0.024.

In some situations, it is found that assembly times can be reduced by providing the operators with oriented parts from automatic feeding and orienting devices. In this case, the operator time is reduced to t_0' and the equations for production time and cost of assembly become

$$t_{pr} = kt_0'(1 + x) \tag{7.49}$$

$$C_{pr} = \frac{nWt_{pr}}{k}\left(1 + \frac{2C_B + C_c + kC_F}{SQ}\right) \tag{7.50}$$

If, in this situation, the operator assembly time t_0' is reduced to 6 s, the assembly cost per part becomes

$$\frac{C_{pr}}{n} = 0.0143 + 0.001k \tag{7.51}$$

Equating this result to the assembly cost of $0.024 per part obtained for the system without feeders, it is found that a break-even value of k is around 10. This means that, for the figures used, when each operator is assembling less than 10 parts, it will be economical to provide feeding and orienting devices.

Of course, the number of parts assembled by each operator affects the production rate, and the results of this analysis are better illustrated on a plot of assembly cost against production rate. A convenient measure of production rate is the annual production volume V_a. Normally, there are 2000 hours available per shift in one year. To allow for operator breaks and machine breakdowns (except those due to faulty parts), it will be assumed that 70% of this time is actually spent working. Thus,

$$V_a = \frac{S(2000)(0.7)(3600)}{t_{pr}}$$

$$= \frac{S}{t_{pr}} (5.04 \times 10^6) \tag{7.52}$$

when the production time t_{pr} is given in seconds. The results for multistation operator assembly are now plotted in Fig. 7.9 and show that for a production volume greater than 150,000 per year, the provision of feeders on the assembly line is economical.

Fig. 7.9 Relation between cost of assembly and annual production volume of operator assembly lines.

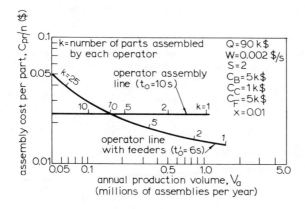

7.7 Single-Station Operator Assembly

When the number of parts in the assembly is reasonably small, it is
sometimes more economical to have each operator assemble the complete
product. This system is more flexible than the assembly line system
because the production volume can be varied readily by removing or
adding operators as required. The degree of mechanical assistance
with which an operator is provided can vary widely. In some cases,
the operator might simply have a powered screwdriver to help with
screwing operations. Or the other extreme might involve a sophisticated workstation with a variety of mechanical aids operated by foot
pedals. Also, the parts might be presented to the operator by means
of automatic feeding and orienting devices. The idea behind these
mechanical aids is to reduce assembly time and increase the productivity
of the operator. This is an important type of assembly system.

First, it will be assumed that an operator is provided with a work-
table (cost C_s) and boxes of parts. Again assuming that each faulty
part results in a repetition of that particular operation, the mean
production time per assembly is given by

$$t_{pr} = nt_0(1 + x) \tag{7.53}$$

and the assembly cost is

$$C_{pr} = Wt_{pr}\left(1 + \frac{C_s}{SQ}\right) \tag{7.54}$$

For a value of C_s of 10 k\$ and the remaining figures as before, the
assembly cost per part C_{pr}/n is \$0.0236. Assuming next that the
operator is provided with oriented parts, thus reducing the assembly
time per part to t_0', the equations become

$$t_{pr} = nt_0'(1 + x) \tag{7.55}$$

and

$$C_{pr} = Wt_{pr}\left(1 + \frac{C_s + nC_F}{SQ}\right) \tag{7.56}$$

The assembly cost per part for a value of t_0' of 6 s is then

$$\frac{C_{pr}}{n} = 0.0141 + 0.001n \tag{7.57}$$

The break-even value for n is found by equating the cost per part
for the two schemes. For the figures used, n is 10 and means that

under these circumstances, the use of feeders to aid the operator is economical only when the assembly consists of less than 10 parts.

7.8 Programmable Assembly Automation

The possible applications of programmable equipment to automatic assembly will not be studied. These applications can be divided into three areas:

1. Applications of computer control to dedicated multistation machines to allow for variations in product style
2. Applications of programmable workheads (or robots) to multistation machines so that more than one assembly operation can be carried out at each station
3. Completely programmable assembly centers to accommodate product changes

7.8.1 Computer Control of Multistation Machines

When a product is required in large quantities but is subject to style variations, multistation machines can be employed where the individual stations, although dedicated to one task, can be commanded by a computer to carry out that task if required. For example, suppose that a product can have either a yellow or a red cap. In this case, two workstations can be provided, each with its own workhead and feeding device, one for yellow caps and one for red caps; the computer can trigger the operation desired. If this kind of option is repeated for other components, then with only a few alternatives, a large number of product styles can be produced.

The analysis of such a system is basically the same as that for simple dedicated machines, except that the cost of the extra workheads and computer must be included. The principal disadvantage remains that production volumes on the order of 1 million per year are necessary, and since this is still a dedicated machine, the product must be of stable design for several years.

7.8.2 Multistation Machines with Programmable Workheads

These machines are basically dedicated machines, where the single-task workheads are replaced by programmable workheads capable of performing more than one assembly task. This scheme provides flexibility in that the machine can be designed to match more closely the output volume required. In addition, product style changes can be accommodated since the workheads are under computer control and can be commanded to select parts among alternatives available at a particular station.

Programmable workheads can be applied to indexing machines, free-transfer machines, or hybrid machines. It will be assumed in the analysis that the cost of the workheads (or robots) is given by

$$C_{dA} = 25 + 8d \text{ k\$} \tag{7.58}$$

where d is the number of degrees of freedom available and that the cost of the gripper for each workhead is C_g per part to be handled.

It will also be assumed that if the workhead is to handle only one or two parts, only two degrees of freedom are required, for from three to six parts, the number of degrees of freedom will be equal to the number of parts, and for more than six parts, six degrees of freedom will be required.

Indexing Machine. If k parts are assembled at each station, the number of stations is given by n/k. Also, if it is assumed that, because of the generally lower production rate, hand-loaded magazines (cost C_M) are employed instead of parts feeders, then

$$t_{pr} = kt + nxT \tag{7.59}$$

and

$$C_{pr} = t_{pr}[W_t + \frac{n}{k} \frac{W}{SQ} (C_{dA} + C_T + C_c) + \frac{nW}{SQ} (C_g + C_M)] \tag{7.60}$$

where t is the time for each assembly task.

Free-Transfer Machine. In this case, each workhead is connected by buffer storage having a capacity equal to the ratio T/t. Thus,

$$t_{pr} = k(t + xT) \tag{7.61}$$

$$C_{pr} = t_{pr}\{W_t + \frac{n}{k} \frac{W}{SQ} [C_{dA} + (1 + \frac{T}{t})(C_B + \frac{C_c}{2})] + \frac{nW}{SQ} (C_g + C_M)\}$$

$$\tag{7.62}$$

Hybrid Machine. If each individual indexing unit connected by buffer storage has i stations, then

$$t_{pr} = kt + ixT \tag{7.63}$$

and

$$C_{pr} = t_{pr}[W_t + \frac{n}{k}\frac{W}{SQ}(C_{dA} + C_T + C_c) + \frac{n}{ki}\frac{W}{SQ}\frac{T}{t}(C_B + \frac{C_c}{2})$$

$$+ \frac{nW}{SQ}(C_g + C_M)] \qquad (7.64)$$

An optimum indexing unit size will arise and is given by

$$i_{opt}^2 = \frac{(1/x)(C_B + C_c/2)}{(SQW_t/nW) + (C_{dA} + C_T + C_c)/k + C_g + C_M} \qquad (7.65)$$

Assuming values of C_g of 0.5 k\$, C_M of 0.5 k\$, C_{dA} given by Eq. (7.58), $S = 3$, and the remaining values as before, Eq. (7.65) becomes

$$i_{opt}^2 = \frac{550}{270/n + 36/k + 9} \qquad (7.66)$$

when $2 \le k \le 6$. Analysis of Eq. (7.66) indicates that the value of i_{opt} varies between 3 and 6 as the value of n varies between 10 and 50. However, since i (the number of indexing units) cannot be greater than n/k (the number of stations), the possible configurations approaching the optimum are severely limited. In fact, unless the value of k is small, the configuration becomes identical to a single indexing machine. It appears, therefore, that the hybrid system with programmable workheads affords little additional flexibility compared with the free-transfer machine with programmable workheads and will not be considered further.

Figure 7.10 presents the results for the indexing and free-transfer machines [Eqs. (7.60) and (7.61), respectively] for n = 10 and 50 and for three shifts working. It can be seen that when n is small, the indexing machine is the more economical and that when n is large and k is small, the free-transfer machine would be employed. However, it should be pointed out that when k is small, the production volume would be high and the dedicated free-transfer machine might be preferred. It is unlikely that systems such as these will be employed for single product designs because this would not take full advantage of the programmability of the workheads. This programmability feature can be employed in two ways:

1. To accommodate small changes in product design without altering the design of the workhead
2. To allow for product style variations without the need for separate workheads for each alternative part

Taking these features into account will allow economical comparison to be made with simple dedicated multistation machines. For the

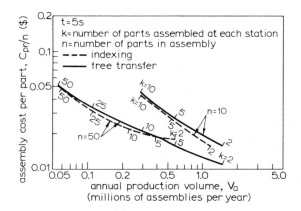

Fig. 7.10 Economics of multistation machines with programmable workheads. C_T = 10 k\$; C_c = 1 k\$; C_g = 0.5 k\$; C_M = 0.5 k\$; C_B = 5 k\$; n = 0.01; T = 30 s; C_{dA} = 25 + 8d k\$; d = k for 2 ≤ k ≤ 6; d = 6 for k > 6; Q = 30 k\$; W = 0.002 \$/s; W_t = 0.006 \$/s; S = 3; x = 0.01.

purposes of these initial comparisons, only the machines using the indexing transfer systems will be studied. Also, to make the comparisons fair, only products consisting of 10 parts will be considered and it will be assumed that feeders are employed for the programmable workheads.

Effect of Changes in Product Design. If N_d is the number of changes in product design anticipated during the life of the machine and each product design change involves only one part requiring one new workhead and feeder on the multistation indexing machine, then Eqs. (7.4) and (7.40) become

$$t_{pr} = t + nxT \qquad (7.67)$$

$$C_{pr} = t_{pr} \left[W_t + \frac{nW}{SQ} (C_T + C_c) + (N_d + n) \frac{W}{SQ} (C_W + C_F) \right] \qquad (7.68)$$

remembering that the costs of the items include design, engineering, and debugging.

For a multistation machine with programmable workheads, Eqs. (7.59) and (7.60) become

$$t_{pr} = kt + nxT \qquad (7.69)$$

$$C_{pr} = t_{pr}[W_t + \frac{n}{k}\frac{W}{SQ}(C_{dA} + C_T + C_c) + \frac{(N_d + n)W}{SQ}(C_g + C_F)]$$

(7.70)

and in this case it is assumed that a change of one part involves a new feeder (cost C_F) and a change to the gripper (cost C_g). Substitution of the values of n = 10, t = 5 s, T = 30 s, S = 3, k = 1, and the remaining numerical values as before, the foregoing equations become:

1. With dedicated workheads:

$$t_{pr} = 8 \text{ s}$$

(7.71)

$$C_{pr} = 0.0942 + 0.0027N_d \quad \$$$

(7.72)

2. With programmable workheads:

$$t_{pr} = 8 \text{ s}$$

(7.73)

$$C_{pr} = 0.15 + 0.001N_d \quad \$$$

(7.74)

These results are presented in Fig. 7.11, where it can be seen that a break-even value of N_d = 33 arises. If more than 33 product design changes are envisaged, the machine with programmable workheads will be more economical.

Effect of Product Style Variations. It was mentioned earlier that a further advantage of the machine with programmable workheads is the ease with which product style variations can be accommodated. On a machine with dedicated workheads, a separate station, workhead, and feeder must generally be provided for each alternative part. For simplicity it is assumed that differing product styles are to be developed by having yn parts available, only n of which are to be assembled in one product, and then the equations for the indexing machines are:

1. With dedicated workheads:

$$t_{pr} = t + nxT$$

(7.75)

$$C_{pr} = t_{pr}[W_t + \frac{ynW}{SQ}(C_T + C_c + C_W + C_F)]$$

(7.76)

2. With programmable workheads:

$$t_{pr} = kt + nxT$$

(7.77)

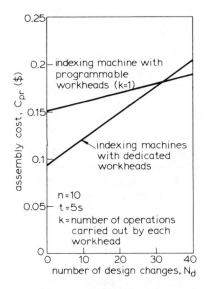

Fig. 7.11 Effect of number of product design changes on the economics of programmable workheads (costs as in Fig. 7.10).

Fig. 7.12 Effect of product style variations on economics of programmable workheads (costs as in Fig. 7.10).

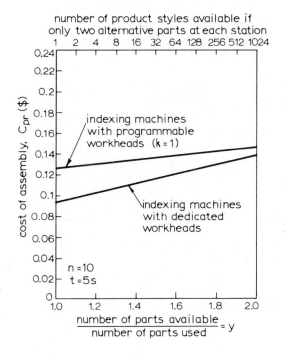

$$C_{pr} = t_{pr}[\ W_t + \frac{n}{k}\frac{W}{SQ}\ (C_{dA} + C_T + C_c) + \frac{ynW}{SQ}\ (C_g + C_F)]$$

(7.78)

The results are plotted in Fig. 7.12, using the same numerical values as before. In this example, even if more than 20 parts are available for a product consisting of 10 parts, the indexing machine with dedicated workheads is still more economical. However, if lower production rates are required, the dedicated machines rapidly become uneconomic, as will be seen later.

7.9 Single-Station Assembly Center

The final system to be considered here is the single-station assembly center, where one sophisticated robot having one or two arms is employed to build an assembly from a complete set of parts. This system provides perhaps the greatest potential for batch assembly work, where batches of differing products are to be assembled or where large numbers of product styles are required. Initially, however, the following questions should be answered:

1. What approach should be used in gripper design?
2. What approach should be used for parts presentation?

7.9.1 Gripper Design

Although sophisticated robots are available which can be quickly programmed to perform complicated series of manipulations, the gripper fixed to the end of the robot arm is usually very limited in its capabilities and often, in fact, must be tailor-made for a particular part. To consider the economics of the various solutions to this problem, a 6-degree-of-freedom robot with one arm will be analyzed. The first possible solution is to provide special individual grippers arranged in racks within reach of the robot arm. Before performing an assembly operation, the robot must replace the previously used gripper in the rack and select the new gripper. Clearly, the time for this manipulation can easily be on the order of the time taken for the assembly operation. This will not only reduce the production time but will increase the cost of assembly.

Assuming that the gripper change time is t_g, the proportion of parts for which a gripper change is necessary is q, and the mean time required to perform an assembly operation is t, then the mean production time is given by

$$t_{pr} = n(t + qt_g + xT)$$

(7.79)

where n is the number of parts in the assembly. The cost of assembly is

$$C_{pr} = t_{pr}[W_t + \frac{W}{SQ}(C_{6A} + C_c) + \frac{nW}{SQ}(C_g + C_M)] \qquad (7.80)$$

where C_{6A} is the cost of the robot and C_M is the cost of the hand-loaded magazines employed as parts presenters. If the value of W_t is again assumed to be 0.006 \$/s, this will allow for one engineer at 0.004 \$/s to supervise the machine and correct faults, together with one assembly operator at 0.002 \$/s to load the magazines. The remaining values are as follows:

Mean assembly time t = 5 s
Time for gripper change t_g = 5 s
Proportion of defective parts x = 0.01
Downtime caused by defective part T = 30 s
Cost of robot with 6 degrees of freedom C_{6A} = 73 k\$
Cost of work carrier C_c = 1 k\$
Cost of gripper for one part only C_g = 1 k\$
Cost of special magazine for each part C_M = 0.5 k\$
Assembly operator rate W = 0.002 \$/s
Operator equivalent in capital equipment Q = 30 k\$
Number of shifts S = 3

Substitution of these figures in Eq. (7.80) results in the graph shown in Fig. 7.13. For the case where q = 0 (no gripper changes), the assembly costs per part rise slightly as n increases. This is simply because of the extra magazines required and is a negligible effect. However, when q = 1 (every part needing a different gripper), the effect of the cost of the extra gripper is shown basically by the rise in cost as n increases. However, the fact that the whole curve is much higher is due mainly to the nonproductive time on the machine due to gripper changes. It is the latter factor that is the biggest problem in employing schemes where gripper changes are necessary.

It is now interesting to speculate on the effect that development of a special gripper that will handle all the parts in an assembly would have on the economics of an assembly center. If the cost of such a gripper is denoted by C_{sg}, the equations for production time and assembly cost become

$$t_{pr} = n(t + xT) \qquad (7.81)$$

and

$$C_{pr} = t_{pr}\left[W_t + \frac{W}{SQ}(C_{6A} + C_c + C_{sg}) + \frac{nWC_M}{SQ}\right] \qquad (7.82)$$

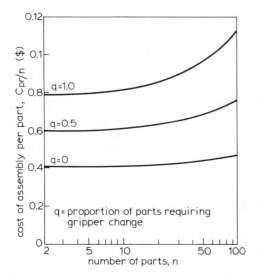

Fig. 7.13 Effect of gripper changes on assembly costs. $x = 0.01$; $T = 30$ s; $C_{dA} = 73$ k$; $S = 3$; $Q = 30$ k$; $W_t = 0.006$ $/s; $W = 0.002$ $/s; $t = 5$ s; $C_c = 1$ k$; $C_g = 0.5$ k$; $C_M = 0.5$ k$.

The economic cost of the special gripper can now be obtained by equating Eqs. (7.80) and (7.82). After rearrangement and assuming that xT is small compared with t,

$$C_{sg} \simeq q \, \frac{t_g C_{TOT}}{t} + nq C_g \qquad (7.83)$$

where C_{TOT} is the total equivalent capital equipment cost of the machine, operators, and individual grippers. It is reasonable to assume that if $t_g = t$ and if $q = 0.5$ (a typical value), an economic cost for a gripper for the machine considered would be 182 k$. The magnitude of this sum indicates that schemes involving gripper changes causing nonproductive time must be avoided.

In this comparison, no account was taken of the possibility of product changes, and therefore the use of a universal (or programmable) gripper should be considered under these circumstances.

Denoting the cost of a universal gripper as C_{ug} and N_p as the number of different products to be assembled during the life of the machine,

$$C_{pr} = t_{pr} \left[W_t + \frac{W}{SQ} (C_{6A} + N_p C_c + C_{ug}) + \frac{N_p nW}{SQ} C_M \right] \qquad (7.84)$$

where t_{pr} is given by Eq. (7.81). With a special gripper for each product, Eq. (7.82) becomes

$$C_{pr} = t_{pr}\left[W_t + \frac{W}{SQ}(C_{6A} + N_pC_c + N_pC_{sg}) + \frac{N_pnW}{SQ}C_M\right] \quad (7.85)$$

where again t_{pr} is given by Eq. (7.81). The economic cost of the universal gripper is found, by equating Eqs. (7.84) and (7.85), to be

$$C_{ug} = N_pC_{sg} \quad (7.86)$$

Clearly, when several product changes are envisaged, the economic value of a universal gripper will be enormous. This is obviously an area that will need to be given special attention by those interested in truly programmable assembly centers.

For a study of parts presentation techniques it will be assumed, as before, that a special gripper for the product is employed and that its cost C_{sg} is C_g per part to be handled, that is,

$$C_{sg} = nC_g \quad (7.87)$$

7.9.2 Parts Presentation Techniques

It is likely that in any assembly center for batch production, the means of part presentation for a given product will form a combination of:

1. Operator loading of large parts and some medium-size parts
2. Special magazines containing medium-size parts and some small parts
3. Special feeders for some medium-size and small parts
4. Adjustable or programmable feeders for some small parts
5. Pallets of prepositioned medium-size and large parts

If p is the proportion of parts presentation stations where a programmable feeder might be used, and if it is assumed that special feeders are employed at the remaining stations, the assembly cost for a machine with one arm and employing a special gripper is

$$C_{pr} = t_{pr}[W_t + \frac{W}{SQ}(C_{6A} + C_c + C_{sg}) + \frac{Wn}{SQ}pC_{PF}$$

$$+ \frac{WN_p}{SQ}n(1 - p)C_F] \quad (7.88)$$

where C_{PF} is the cost of a programmable feeder, N_p the number of product changes, and t_{pr} is given by Eq. (7.81). Differentiation of Eq. (7.88), with respect to p, gives, for a break-even value of C_{PF},

$$C_{PF} = N_p C_F \qquad\qquad (7.89)$$

Thus, as would be expected, the economic cost of a programmable feeder is the cost of a special feeder multiplied by the number of products to be assembled.

Another alternative is the use of special magazines or chutes similar to those employed in some vending machines. Such magazines could be loaded with parts at regular intervals by the machine operator or kept full by means of parts feeders. If the machine operator takes a time t_l to load one part into a magazine, the cost of loading each part will be Wt_l, where W is the operator rate. If the cost of a special feeder is C_F and the number of product changes is N_p, then the cost to handle each part with a feeder will be $N_p W C_F t_{pr}/SQ$, where t_{pr} is given by Eq. (7.81). Thus, by equating these costs, the economic cost of a special feeder can be found to be

$$C_F = \frac{t_l SQ}{N_p n(t + xT)} \qquad\qquad (7.90)$$

Assuming that $t_l = 1$ s, $t = 5$ s, $S = 2$, $Q = 30$ k\$, $x = 0.01$, and $T = 30$ s, then

$$C_F = \frac{11.3}{N_p n} \quad k\$ \qquad\qquad (7.91)$$

When an assembly consists of 10 parts (n = 10) and even if there are no product changes, the economic cost of a special feeder would be only about \$1000. Since the cost of special feeders is presently on the order of \$5000, they will rarely be economical for this application. For a programmable feeder, combining Eqs. (7.89) and (7.91) gives an economic cost of 11.3/n k\$ and the same conclusions can be drawn. Thus, it may be assumed that the use of feeders to load magazines on-line should not be considered. Also, since the delivery track for a feeder can be regarded as a magazine, conventional feeders with delivery tracks, either of the special or programmable kind, will not generally be economical for programmable assembly automation.

In summary, bearing in mind that inexpensive programmable feeders are not presently available, it is reasonable to assume that special magazines for each part (cost C_M), hand-loaded on-line, will be used for programmable assembly automation.

7.9.3 Effect of Parts Quality

The mean production time for an assembly center using a robot with one arm is given by

$$t_{pr} = n(t + xT) \tag{7.92}$$

The proportion of downtime D is therefore

$$D = \frac{\text{downtime}}{\text{production time}}$$

$$= \frac{nxT}{n(t + xT)}$$

$$= \frac{x}{x + t/T} \tag{7.93}$$

If $x = 0.01$ and $t/T = 0.2$, as before, then D is 0.05, which, for large values of n, is an order of magnitude less than the figure obtained for multistation indexing machines. Thus, it can be concluded that, with assembly centers, the downtime due to defective parts will generally not be a severe problem.

7.9.4 Assembly Centers Using Robots with Two Arms

Part of the time taken to perform an assembly operation is used in selecting a part from a magazine or feeder and transporting it to the assembly area. By utilizing two arms on an assembly center, a reduction in the mean assembly time t_{pr} can be achieved whereby one arm is selecting the next part while the other arm is performing the assembly operation with the preceding part. If it can be assumed that the operations of one arm never interfere with the other, the mean assembly time is given by

$$t_{pr} = n(\frac{t}{2} + xT) \tag{7.94}$$

and the assembly cost will be

$$C_{pr} = t_{pr} \{ W_t + \frac{W}{SQ} [2C_{dA} + N_p C_c + N_p n(C_g + C_M)] \} \tag{7.95}$$

assuming hand-loaded magazines and a special gripper for each product. This equation is compared with the corresponding equation [Eq. (7.85)] for an assembly center employing a robot with one arm in Fig. 7.14. It can be seen that, with the figures used and assumptions made, an assembly center with two robot arms will always be significantly more economical than an assembly center using one robot arm. However, it can be seen that assembly costs per part increase rapidly when the number of parts in the assembly is greater than about 10 and when a

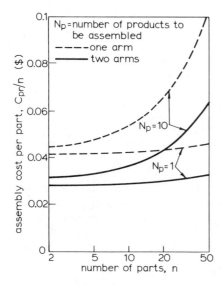

Fig. 7.14 Comparison of assembly costs for assembly centers employing one robot arm and two robot arms.

significant number of different products are to be assembled. Clearly, if various products are to be assembled and unless the required production rate is very low, it would be necessary to consider dividing the product into subassemblies and employing a series of assembly centers connected by buffer storage.

7.10 Comparison of Different Assembly Systems

Economic comparisons will now be made between the various systems described above. It should be emphasized that such comparisons depend heavily on the assumptions made and cost figures employed. It is felt, however, that certain general conclusions can be made regarding the assembly system most likely to be economical for a given application. In order to reduce the number of alternative schemes presented in each figure, comparisons will only be made among dedicated assembly automation, programmable automation, and manual assembly. When presenting data for these categories, the results for the best system in each category will be chosen. For example, in Fig. 7.8, a comparison was made between three types of dedicated automatic assembly equipment. For the present purposes, only one relationship would be presented, that giving the minimum cost for the particular circumstances. For the conditions of Fig. 7.8, this would mean employing an indexing machine when $n \leqslant 8$ and a hybrid machine when $n > 8$.

One additional system is considered which attempts to predict the economics of a machine that might be developed in the future. This system will be termed the universal assembly center and is basically an assembly center with two arms, each having only three degrees of freedom and fitted with universal grippers. Parts presentation is by means of programmable feeding devices.

The cost of each arm is again given by Eq. (7.58), where d is 3, that is, 49 k$, and is a figure that could be achieved with present-day technology. However, the universal gripper cost C_{ug} is assumed to be 10 k$, and the cost for each programmable feeder C_{pf} is 1 k$. These items at these costs do not yet exist, but the results of the comparisons will show the economic benefits that could be gained from their development.

It is not intended that a universal assembly center, if developed, would be capable of assembling existing products. Rather, it is assumed that by close collaboration between design and manufacturing personnel, the machine would be used only for new products that have been designed to be assembled on this particular machine. This means, for example, that assembly operations are all carried out from above in one direction only and can be accomplished by a robot arm having only three degrees of freedom. Also, it is clear that a universal gripper and programmable feeding devices will never be developed that will handle all possible parts. It is therefore assumed again that parts will be employed that are capable of being handled by the equipment developed. With these assumptions, the equations for a universal assembly center are

$$t_{pr} = n(\frac{t}{2} + xT) \qquad (7.96)$$

$$C_{pr} = t_{pr}[W_t + \frac{W}{SQ}(2C_{3A} + ynC_{PF} + 2C_{ug} + N_pC_c)] \qquad (7.97)$$

Also, since an operator is not required to load magazines, the value of W_t is assumed to be 0.004 $/s, representing one supervising engineer.

Figure 7.15 shows diagramatically all the representative assembly systems compared here. In Figs. 7.16 through 7.23, the assembly cost per part C_{pr}/n is plotted against the annual production volume V_a for three-shift working. The individual effects of the following factors are illustrated:

1. The number of parts in the product, n
2. The stability of the product measured by the number of design changes anticipated during the life of the machine, N_d

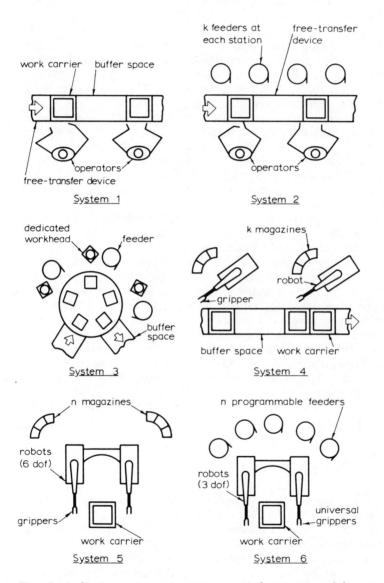

Fig. 7.15 Various assembly systems (dof, degrees of freedom).

Table 7.4 Summary of Equations Employed in Comparison of Assembly Systems

Assembly system	t_{pr}	W_t	M'
1. Operator assembly line, no feeders	$kt_0(1 + x)$	$\dfrac{nW}{k}$	$(n/k)(2C_B + N_pC_c)$
2. Operator assembly line, with feeders	$kt_0'(1 + x)$	$\dfrac{nW}{k}$	$(n/k)(2C_B + N_pC_c) + N_p(ny + N_d)C_F$
3. Dedicated hybrid machine	$t + xiT$	$3W$	$[nyC_T + (T/t)(ny/i)C_B]$ $+ N_p\{(ny + N_d)(C_F + C_W) + [ny + (T/2t)(ny/i)]C_c\}$
4. Free-transfer machine with programmable workheads	$k(t + xT)$	$3W$	$(n/k)[C_{dA} + (T/t + 1)C_B]$ $+ N_p[(ny + N_d)C_M + nC_g + (n/k)(T/2t + 0.5)C_c]$
5. Assembly center with two arms	$n(t/2 + xT)$	$3W$	$2C_{6A} + N_p[C_c + nC_g + (ny + N_d)C_M]$
6. Universal assembly center	$n(t/2 + xT)$	$2W$	$(2C_{3A} + nyC_{PF} + 2C_{ug}) + N_pC_c$

Notes:

$C_{pr} = t_{pr}(W_t + WM'/SQ)$

$i_{opt}^2 = (nyC_B + \tfrac{1}{2}nyN_pC_c)/x[3SQ + ny(C_T + N_pC_c) + (ny + N_d)N_p(C_F + C_W)]$

$d = 2$ if $k < 2$; $d = k$ if $2 \leq k \leq 6$; $d = 6$ if $k > 6$

d is the degree of freedom and k is the number of parts assembled by each operator or programmable workhead

Table 7.5 Summary of Data Employed in Comparison of Assembly Systems

Description	Symbol
Cost of programmable robot or workhead	$C_{dA} = 25 + 8d$ k$
Cost of transfer device per work station or buffer space on free-transfer machine	$C_B = 5$ k$
Cost of work carrier	$C_c = 1$ k$
Cost of automatic feeding device and delivery track	$C_F = 5$ k$
Cost of gripper per part to be handled	$C_g = 0.5$ k$
Cost of manually loaded magazine	$C_M = 0.5$ k$
Cost for one programmable feeder	$C_{PF} = 1$ k$
Cost of workstation for single-station assembly	$C_s = 10$ k$
Cost of transfer device per workstation for an indexing machine	$C_T = 10$ k$
Cost of universal gripper	$C_{ug} = 10$ k$
Cost of dedicated workhead	$C_W = 10$ k$
Equivalent cost of operator in terms of capital equivalent	$Q = 30$ k$
Number of shifts	$S = 3$
Machine downtime due to defective parts	$T = 30$ s
Operator downtime due to defective parts	t_0 or t_0'
Mean time of assembly for one part	$t_0 = 10$ s operator, no feeder
	$t_0' = 6$ s operator, with feeders
	$t = 5$ s machines
Operator's rate	$W = 0.002$ $/s
Ratio of faulty parts to acceptable parts	$x = 0.01$

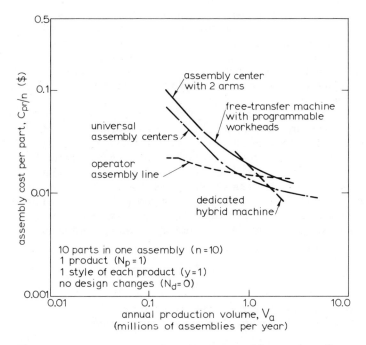

Fig. 7.16 Comparison of various assembly systems for one product having 10 parts.

3. The number of different products to be assembled during the life of the machine, N_p
4. The number of different styles of the same product measured by y, which is the ratio of the number of parts available to the number of parts actually used in one product

The equations developed earlier have been modified to include these effects, and the modified versions are presented in Table 7.4. Table 7.5 presents a summary of the values of the various parameters employed in the comparisons.

Figure 7.16 first shows how the cost per part C_{pr}/n varies with the annual production volume for the four assembly systems when the product contains 10 parts, when only one product with one style is to be assembled, and when there are no design changes. Ignoring for the moment the hypothetical universal assembly centers, it can be seen that for large production volumes, the dedicated hybrid machine is the most economical for large production volumes (more than 1 million per year) and that for all other production volumes, the operator assembly line is the most economical. This clearly represents the present situation. It should be pointed out that the variations in production volume for the dedicated machine have been obtained by varying the cycle time t_{pr} from 5 s to 20 s, whereas for the free-transfer machine with programmable workheads, an assembly time of 5 s has been assumed and the production volume has been varied by allowing each workhead to perform various numbers of tasks (k has been varied). The production volume for the operator assembly line has also been varied by allowing the operators different numbers of assembly tasks. The general characteristics of these curves are similar to the results presented by Lynch,* who in his analysis did not consider the effects of downtime due to defective parts. The use of hypothetical assembly centers (each one assembling the complete product), where sufficient centers are employed to meet the required production volume, shows a small economic advantage over the programmable machines using present-day technology (hand-loaded magazines

---

*Ibid.

Fig. 7.17 Comparison of various assembly systems for a product having 10 parts when 20 design changes are anticipated.

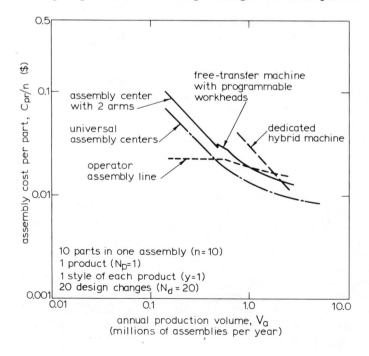

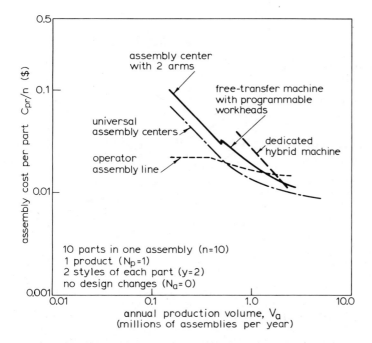

Fig. 7.18 Comparison of various assembly systems for a product having 10 parts with different styles.

and special grippers for the product). However, as would be expected, programmable automation of any kind does not seem really warranted in the situation depicted in this graph.

Figures 7.17 and 7.18 show the effects of design changes (N_d = 20) and product style variations (y = 2), respectively. These effects are similar and indicate that for high production volumes, the programmable equipment using present-day technology becomes competitive. For medium and low production volumes, however, the operator assembly line is the most economic. The hypothetical universal assembly center would be most economical for both medium and high production volumes (above 500,000 per year).

Figure 7.19, where the effect of assembling 20 different products is illustrated, shows more startling results. It can be seen that the dedicated machine, as would be expected, is completely uneconomical. However, the programmable free-transfer machine has also become uneconomic, whereas the hypothetical universal universal assembly center shows considerable savings compared with operator assembly at medium and high production volumes.

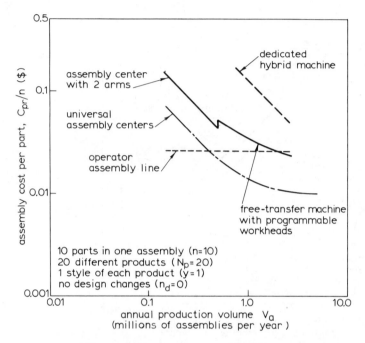

Fig. 7.19 Comparison of various assembly systems for 20 different products each having 10 parts.

Figures 7.20 through 7.23 present the same comparisons as the previous figures, but in these cases the product contains 50 parts instead of 10. The results are similar except that the free-transfer machine with programmable workheads is competitive for medium-volume production when one product is to be assembled and when design changes or different product styles are involved. From all the foregoing comparisons, the following conclusions can be drawn.

1. Dedicated machines should be considered when the production volume is high and when one product with one style is to be assembled. However, for high production volumes but where design changes and product style variations are anticipated, the dedicated machines might still be the most economical, especially for products having large numbers of parts.
2. Free-transfer machines with programmable workheads are marginally economical for products containing few parts when the production volume is high and when product design changes and various product styles are involved. However, when the product contains a large

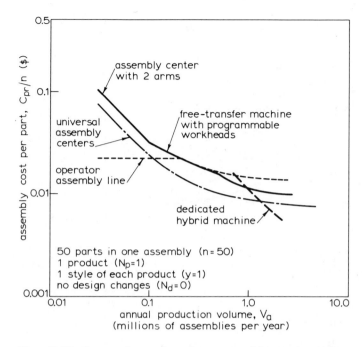

Fig. 7.20 Comparison of various assembly systems for one product having 50 parts.

Fig. 7.21 Comparison of various assembly systems for a product having 50 parts when 20 design changes are anticipated.

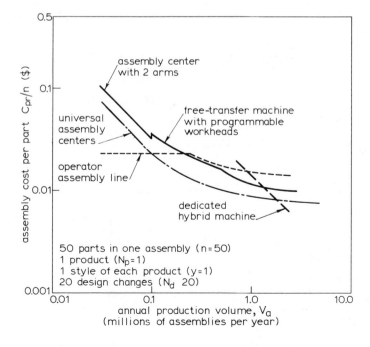

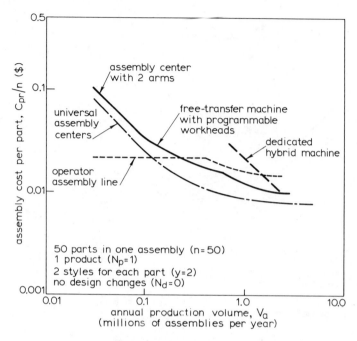

Fig. 7.22 Comparison of various assembly systems for a product having 50 parts with different styles.

Fig. 7.23 Comparison of various assembly systems for 20 different products each having 50 parts.

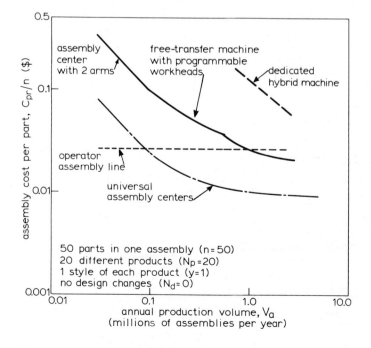

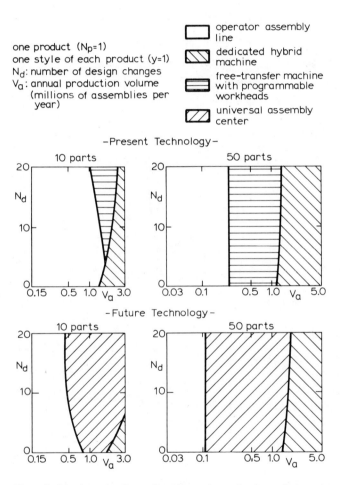

one product (N_p=1)
one style of each product (y=1)
N_d: number of design changes
V_a: annual production volume
   (millions of assemblies per
   year)

☐ operator assembly line

▨ dedicated hybrid machine

▤ free-transfer machine with programmable workheads

▨ universal assembly center

Fig. 7.24 Areas of application when design changes are envisaged.

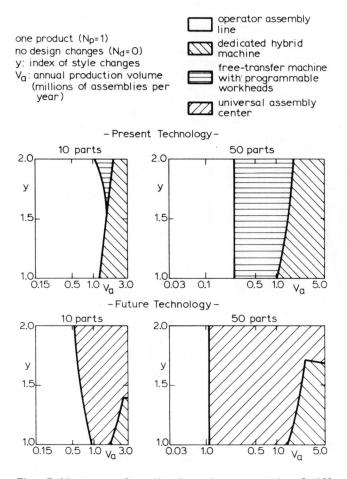

Fig. 7.25 Areas of application when assembly of different product styles is envisaged.

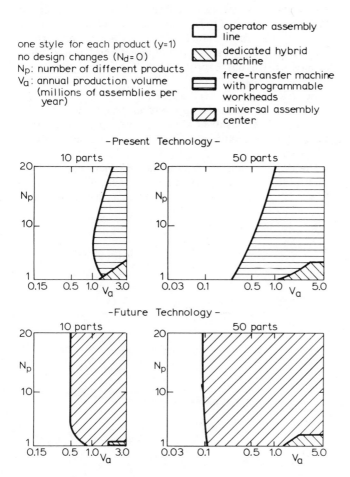

Fig. 7.26 Areas of application when assembly of different products is envisaged.

number of parts, these machines should be considered for medium production volumes when one product is to be assembled and when design changes or particular product style variations are involved. They can also be considered for high production volumes when different products are to be assembled.

3. The hypothetical universal assembly centers, if developed, would change the picture radically. These machines would be competitive for all but very low production volumes (i.e., when at least the full output from one assembly center is required) and would be most economical when different products are to be assembled.

These conclusions are better illustrated in Figs. 7.24 through 7.26, where the various regions of economic application for the various systems are presented. These figures show separately the effects of product design changes, varying product styles, and assembly of different products, respectively.

It is clear that the development of programmable feeders and universal grippers and the design of products for assembly would extend enormously the areas of application of programmable assembly automation.

Chapter 8

Design for Automatic Assembly

Experience shows that it is difficult to make large savings in cost by the introduction of automatic assembly in the manufacture of an existing product. In those cases where large savings are claimed, examination will show that often the savings are really due to changes in the design of the product necessitated by the introduction of the new process. It can probably be stated that, in most of these instances, even greater savings would be made if the new product were to be assembled manually. Undoubtedly, the greatest cost savings are to be made by careful consideration of the design of the product and its individual component parts.

When a product is designed, consideration is generally given to the ease of manufacture of its individual parts and the function and appearance of the final product. Although for obvious reasons it must be possible to assemble the product, little thought is usually given to those aspects of design that will facilitate assembly of the parts and great reliance is often placed on the dexterity of the assembly operators. An operator is able to select, inspect, orient, transfer, place, and assemble the most complicated parts relatively easily, but many of these operations are difficult, if not impossible, to duplicate on a machine. Thus, one of the first steps in the introduction of automation in the assembly process is to reconsider the design of the product so that the individual assembly operations become sufficiently simple for a machine to perform.

The subject of design for automatic assembly can be conveniently divided into two sections: product design for ease of assembly, and design of parts for feeding and orienting.

8.1   Product Design for Ease of Assembly

The most obvious way in which the assembly process can be facilitated
at the design stage is by reducing the number of different parts to a
minimum.   An example of this is given by Iredale* and shown in Fig.
8.1.   Here the original design consisted of 13 parts and required many
difficult operations to assemble.   The new design reduced the product
to two parts, requiring only one simple operation for assembly.   Clearly,
great savings in production cost would have been brought about by
this reconsideration of the design of the product.   Further examples
of product simplification for automatic assembly are given in Fig. 8.2
    It is sometimes possible to simplify a product by employing one of
the new processes that enable complex parts to be produced.   These
parts can sometimes replace complete subassemblies and hence eliminate
many assembly operations.   The new forming and casting operations
are examples of processes that may help in this respect and in particular, precision die casting is being increasingly applied.
    One important factor may arise during the redesign of a product.
For instance, it might be suggested in a particular situation that a
screw, nut, and washer might be replaced by a rivet, or alternatively,
that the parts might be joined by welding or by the use of adhesives.
This would eliminate at least two assembly operations but would result
in a product that would be more difficult to repair.   This illustrates
a common trend, where the introduction of automation may result in a
cheaper product but one that is quite uneconomical to repair.   In the
future, consumers will probably become more accustomed to the idea

---

*R. Iredale, "Automatic Assembly—Components and Products,"
Metalwork. Prod., Apr. 8, 1964.

Fig. 8.1 Reduction of parts to save assembly costs.   (From R. Iredale,
"Automatic Assembly—Components and Products," *Metalwork. Prod.*,
April 8, 1964.)

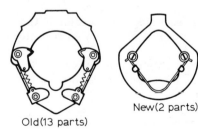

New(2 parts)

Old(13 parts)

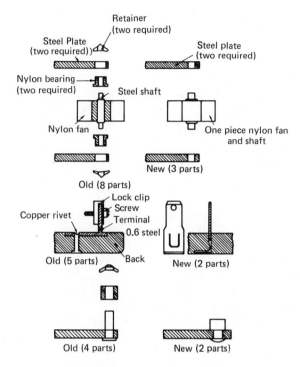

Fig. 8.2 Further examples of product simplification. (From R. Iredale, "Automatic Assembly—Components and Products," *Metalwork. Prod.*, April 8, 1964.)

of replacing the complete product or a major subassembly in the event of a failure.

Apart from product simplification, great improvements can often be made by the introduction of guides and tapers which directly facilitate assembly. Examples of this are given by Baldwin* and Tipping[†] in Figs. 8.3 and 8.4. In both these examples, sharp corners are removed so that the part to be assembled is guided into its correct position during assembly.

---

*S. P. Baldwin, "How to Make Sure of Easy Assembly," *Tool Manuf. Eng.*, May 1966, p. 67.

[†]W. V. Tipping, "Component and Product Design for Mechanized Assembly," Conference on Assembly, Fastening, and Joining Techniques and Equipment, PERA, 1965.

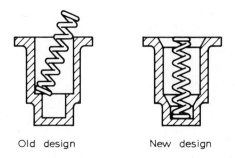

Old design   New design

Fig. 8.3 Redesign of product for ease of assembly. (From S. P. Bladwin, "How to Make Sure of Easy Assembly," *Tool Manuf. Eng.*, May 1966, p. 67.)

Fig. 8.4 Redesign to assist assembly. (From W. V. Tipping, "Component and Product Design for Mechanized Assembly," Conference on Assembly, Fastening, and Joining Techniques and Equipment, PERA, 1965.)

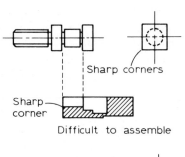

Sharp corners

Sharp corner

Difficult to assemble

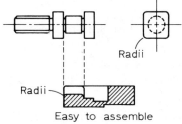

Radii

Radii

Easy to assemble

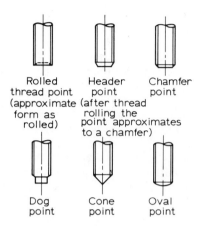

Rolled thread point (approximate form as rolled)   Header point (after thread rolling the point approximates to a chamfer)   Chamfer point

Dog point   Cone point   Oval point

Fig. 8.5 Various forms of screw point.

Further examples in this category can be found in the type of screw used in automatic assembly. Those screws that tend to centralize themselves in the hole will give the best results in automatic assembly and Tipping* summarizes and grades the designs of screw points available as follows (Fig. 8.5):

1. Rolled threat point: very poor location; will not centralize without positive control on the outside diameter of the screws.
2. Header point: only slightly better than (1) if of correct shape.
3. Chamfer point: reasonable to locate.
4. Dog point: reasonable to locate.
5. Cone point: very good to locate.
6. Oval point: very good to locate.

Tipping recommends that only the cone and oval point screws be used in automatic assembly.

Another factor to be considered in design is the difficulty of assembly from directions other than directly above. The aim of the designer should be to allow for assembly in sandwich or layer fashion, each part being placed on top of the previous one. The biggest advantage of this method is that gravity can be used to assist in the feeding and placing of the parts. It is also desirable to have workheads and feeding devices above the assembly station, where they will be accessible in the event of a fault due to the feeding of a defective part.

---

*Ibid.

Assembly from above may also assist in the problem of keeping parts
in their correct positions during the machine index period, when
dynamic forces in the horizontal plane might tend to displace them.
In this case, with proper product design where the parts are self-
locating, the force due to gravity should be sufficient to hold the
part until it is fastened or secured.

If assembly from above is not possible, it is probably wise to divide
the assembly into subassemblies. For example, an exploded view of
a power plug is shown in Fig. 8.6 and in the assembly of this product
it would be relatively difficult to position and drive the two cord grip
screws from below. The remainder of the assembly (apart from the
main holding screw) can be conveniently built into the base from above.
In this example the two screws, the cord grip, and the plug base could
be treated as a subassembly dealt with prior to the main assembly
machine.

It is always necessary in automatic assembly to have a base part
on which the assembly can be built. This base part must have features
that make it suitable for quick and accurate location on the work
carrier. Figure 8.7a shows a base part for which it would be difficult
to design a suitable work carrier. In this case, if a force were applied
at A, the part would rotate unless adequate clamping were provided.
One method of ensuring that a base part is stable is to arrange that

Fig. 8.6 Assembly of three-pin power plug.

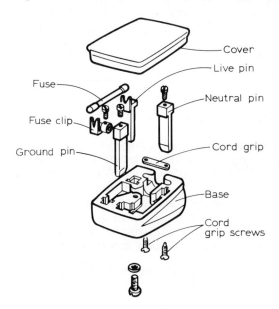

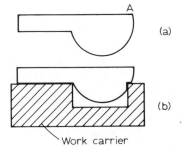

Fig. 8.7 Design of base part for mounting on work carrier.

its center of gravity be contained within flat horizontal surfaces. For example, a small ledge machined into the part will allow a simple and efficient work carrier to be designed (Fig. 8.7b).

Location of the base part in the horizontal plane is often achieved by dowel pins mounted in the work carrier. To simplify the assembly of the base part onto the work carrier, the dowel pins can be tapered to provide guidance, as in the example shown in Fig. 8.8.

Fig. 8.8 The use of tapered pegs to facilitate assembly.

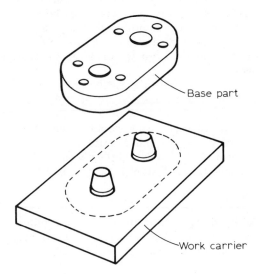

8.2 Design of Parts for Feeding and Orienting

Many types of parts feeders are used in automatic assembly, and some
of them have been studied in an earlier chapter. Most feeders are
suitable for feeding only a very limited range of part shapes and are
not generally relevant when discussing the design of parts for feeding
and orienting. The most versatile parts feeder is the vibratory bowl
feeder, and the following section deals mainly with the aspects of the
design of parts, which will facilitate feeding and orienting in these
devices. Many of the points made, however, apply equally to other
feeding devices.

Much work has been carried out at the University of Massachusetts
at Amherst on the design of small parts for automatic handling.* The
basic idea of this work is to provide designers who are not necessarily
familiar with the problem of automatic handling, with rapid feedback
on their proposed designs of components, indicating the ease or difficulty with which the components could be automatically fed and oriented.
The feedback can also indicate what changes designers might make to
their designs to reduce or eliminate the problems of handling the components. This is done through a coding system specially designed
for the purpose, but similar to some of the coding systems being
developed for various group technology applications. The coding
system developed at the University of Massachusetts at Amherst is
presented in Appendix III and essentially asks a series of questions
regarding the symmetry and geometric or nongeometric features of a
component. Depending on the answers to these questions, a three-
digit code number is assigned to the component. This code number
then indicates the ease or difficulty of feeding and orienting the part
and can be used to obtain information on the handling techniques
available. More important for the present discussion, the process of
obtaining the code number indicates what design features can be changed
to simplify the handling problems.

Taking as an example the long stud shown in Fig. 8.9, which is
threaded unequally at each end, the code number would be 290, indicating a category of parts that are very difficult to handle. If the threaded
portions were made the same length, then the code would become 200.
Parts with this code number are very easy to handle automatically.

Some further examples are shown in Fig. 8.9 together with the
corresponding code numbers. Three basic design principles can be
enumerated:

---

*G. Boothroyd, C. Poli, and L. E. Murch, "The Handbook of Feeding
and Orienting Techniques for Small Parts," Department of Mechanical
Engineering, University of Massachusetts, Amherst, Mass., 1977.

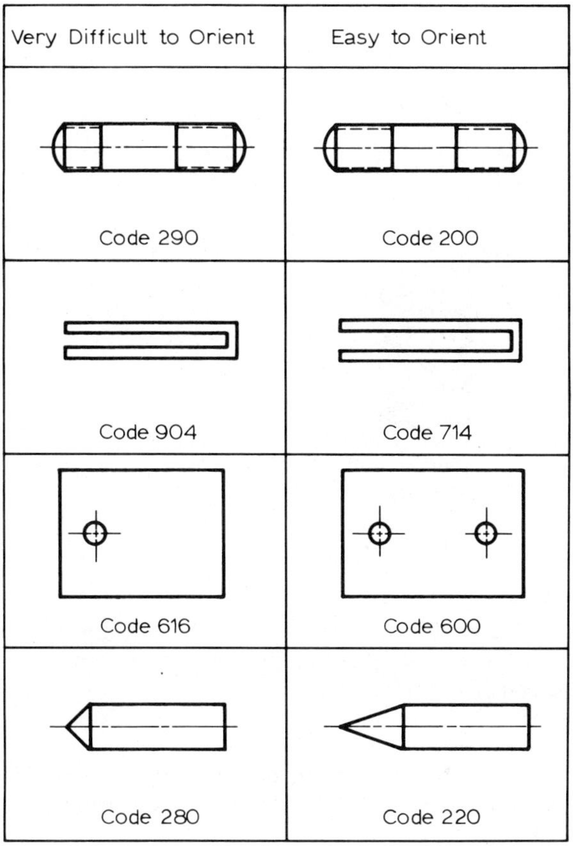

Very Difficult to Orient	Easy to Orient
Code 290	Code 200
Code 904	Code 714
Code 616	Code 600
Code 280	Code 220

Fig. 8.9 Simple design changes to eliminate feeding and orienting problems. (From G. Boothroyd, C. Poli, and L. E. Murch, "The Handbook of Feeding and Orienting Techniques for Small Parts," Department of Mechanical Engineering, University of Massachusetts, Amherst, Mass., 1977, Appendix III.)

1. Avoid designing parts that will tangle or nest.
2. Make the parts symmetrical
3. If parts cannot be made symmetrical, avoid slight asymmetry or asymmetry resulting from small or nongeometrical features.

For parts that tend to tangle or nest when stored in bulk, it can be almost impossible to separate, orient, and feed the parts automat-

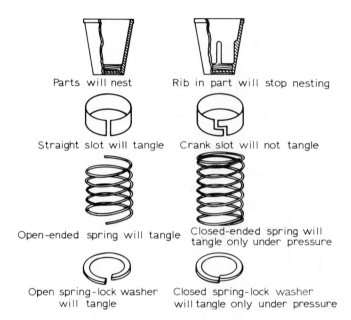

Parts will nest   Rib in part will stop nesting

Straight slot will tangle   Crank slot will not tangle

Open-ended spring will tangle   Closed-ended spring will tangle only under pressure

Open spring-lock washer will tangle   Closed spring-lock washer will tangle only under pressure

Fig. 8.10 Examples of redesign to prevent nesting or tangling of parts. (From R. Iredale, "Automatic Assembly—Components and Products," *Metalwork. Prod.*, April 8, 1964.)

ically. Often a small nonfunctional change in design will prevent this occurrence and some simple examples of this are illustrated in Fig. 8.10.

While the asymmetrical feature of a part might be exaggerated to facilitate orientation, an alternative approach is to deliberately add asymmetrical features for the purpose of orienting. The latter approach is more common and some examples, given by Iredale,* are reproduced in Fig. 8.11. In each case, the features that require alignment are difficult to utilize in an orienting device, so corresponding external features are added deliberately.

It will be noted that in the coding system shown in Appendix III, those parts with a high degree of symmetry all have codes representing parts easy to handle and that all parts with slight asymmetry have codes representing parts very difficult to handle. There are, however, a wide range of codes representing parts that will probably be difficult to handle automatically; these parts will create problems for designers

---

*op. cit.

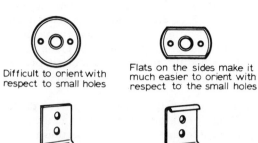

Difficult to orient with respect to small holes

Flats on the sides make it much easier to orient with respect to the small holes

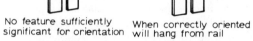

No feature sufficiently significant for orientation

When correctly oriented will hang from rail

Triangular shape of part makes automatic hole orientation difficult

Nonfunctional shoulder permits proper orientation to be established in a vibratory feeder and maintained in transport rails

Fig. 8.11 Provision of asymmetrical features to assist in orientation. (From R. Iredale, "Automatic Assembly—Components and Products," *Metalwork. Prod.*, April 8, 1964.)

Fig. 8.12 Less obvious example of a design change to simplify feeding and orienting problems.

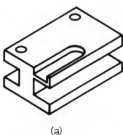

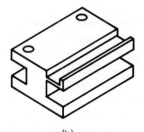

(a)
Very difficult to orient
Code 857

(b)
Possible to orient
Code 843

unless they are provided with assistance. Figure 8.12a shows a part that would be difficult to handle and Fig. 8.12b shows the redesigned part, which could be fed and oriented in a vibratory bowl feeder at a high rate. The subtle change in design would not be obvious to the designer without the use of the coding system. In fact, it probably would not have occurred to the designer that the original part was difficult to handle.

It should be pointed out that, although the discussion above dealt specifically with automatic handling, parts that are easy to handle automatically will also be easy to handle manually. A reduction in the time taken for an assembly operator to recognize the orientation of a part and then reorient it results in considerable cost savings.

Clearly, with some parts, it will not be possible to make design changes that will enable them to be handled automatically: for example, very small parts or complicated shapes formed from thin strip and difficult to handle in an automatic environment. In these cases it is sometimes possible to manufacture the parts on the assembly machine or to separate them from the strip at the moment of assembly. Operations such as spring winding or blanking out thin sections have been successfully introduced on assembly machines in the past.

8.3 Summary of Design Rules

The various points made in this discussion of parts and product design for automatic assembly are summarized below in the form of simple rules for the designer.

8.3.1 Rules for Product Design

1. Minimize the number of parts.
2. Ensure that the product has a suitable base part on which to build the assembly.
3. Ensure that the base part has features that will enable it to be readily located in a stable position in the horizontal plane.
4. If possible, design the product so that it can be built up in layer fashion, each part being assembled from above and positively located so that there is no tendency for it to move under the action of horizontal forces during the machine index period.
5. Try to facilitate assembly by providing chamfers or tapers which will help to guide and position the parts in the correct position.
6. Avoid expensive and time-consuming fastening operations, such as screwing, soldering, and so on.

8.3.2 Rules for the Design of Parts

1. Avoid projections, holes, or slots that will cause tangling with identical parts when placed in bulk in the feeder. This may be achieved by arranging that the holes or slots are smaller than the projections.
2. Attempt to make the parts symmetrical to avoid the need for extra orienting devices and the corresponding loss in feeder efficiency.
3. If symmetry cannot be achieved, exaggerate asymmetrical features to facilitate orienting or, alternatively, provide corresponding asymmetrical features that can be used to orient the parts.

8.4 Analysis of Design for Assembly

A handbook entitled "Design for Assembly" has been developed at the University of Massachusetts at Amherst.* This handbook describes systematic procedures for the analysis of the designs of products or assemblies for either manual assembly or automatic assembly. The handbook is intended to provide a systematic way of examining a complete product or subassembly design, drawing the designer's attention to those part and product design features that are likely to involve high handling and assembly costs. In this way the designer will be encouraged to consider design changes or even alternative designs that will reduce handling and assembly costs.

Before a design can be analyzed it must be decided whether the assembly processes are mainly to be carried out manually or with the aid of automation. For this reason, the handbook is divided into three parts. The first part (sec. 1) contains a classification system for products or assemblies. It can be used to obtain an indication of whether automatic assembly should be considered for a particular product or subassembly (this classification system is briefly described at the end of Chapter 9 of the present text); the second part of the handbook (secs. 2, 3, and 4) is a system for the analysis of product or assembly designs suitable for manual handling and assembly; the third part (secs. 5 and 6) is a system for the analysis of product or assembly designs suitable for automatic assembly.

Analysis of the design of products or assemblies for automatic handling and assembly is basically carried out using two coding systems: (1) automatic handling processes and (2) automatic assembly processes. The three-digit code described in Appendix III is used for handling processes and a two-digit code (a portion of which is illustrated in

---

*G. Boothroyd, "Design for Assembly," Department of Mechanical Engineering, University of Massachusetts, Amherst, Mass., 1979.

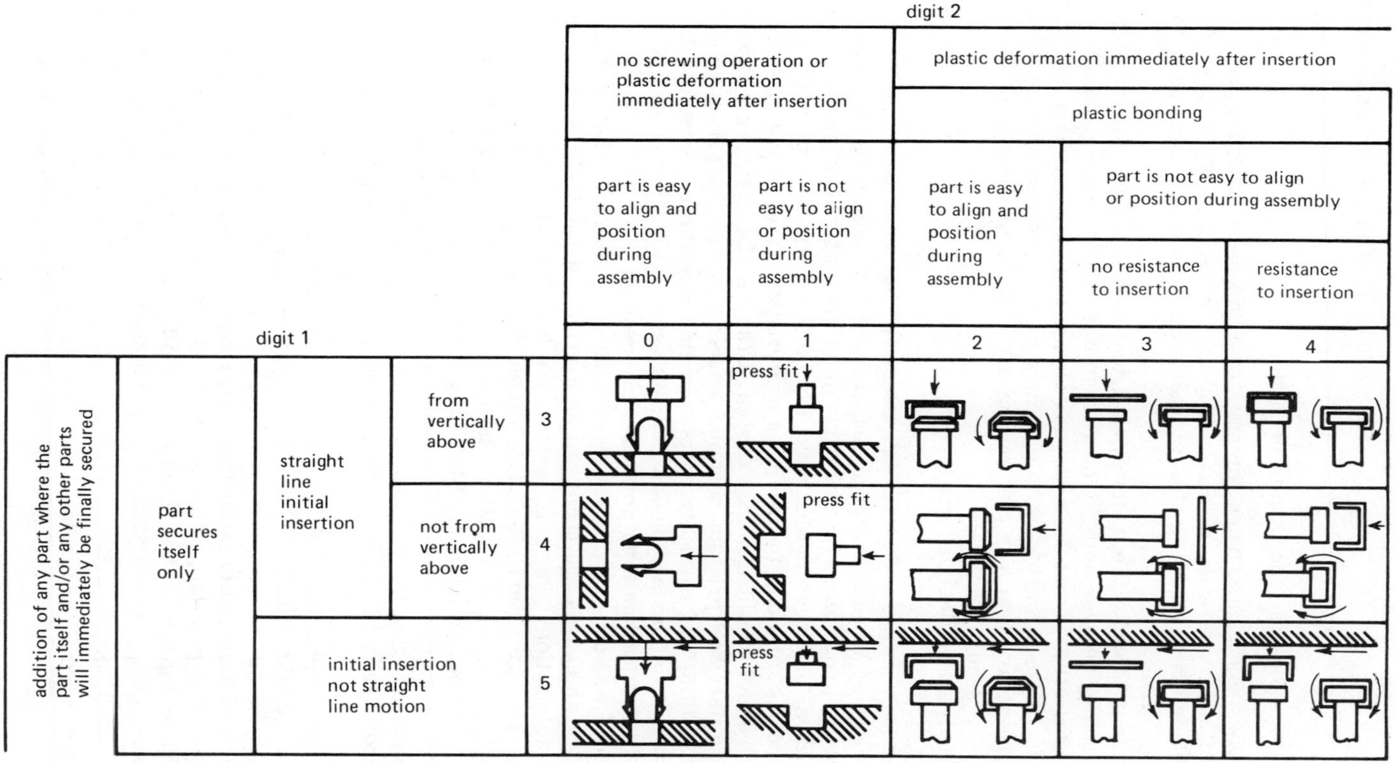

digit 2

		plastic deformation immediately after insertion			
		no screwing operation or plastic deformation immediately after insertion		plastic bonding	
				part is not easy to align or position during assembly	
		part is easy to align and position during assembly	part is not easy to align or position during assembly	part is easy to align and position during assembly	no resistance to insertion	resistance to insertion
		0	1	2	3	4

digit 1

part secures itself only	straight line initial insertion	from vertically above — 3
		not from vertically above — 4
	initial insertion not straight line motion	5

addition of any part where the part itself and/or any other parts will immediately be finally secured

Fig. 8.13 Coding system for automatic assembly processes. (From G. Boothroyd, "Design for Assembly," Department of Mechanical Engineering, University of Massachusetts, Amherst, Mass., 1979.)

press fit

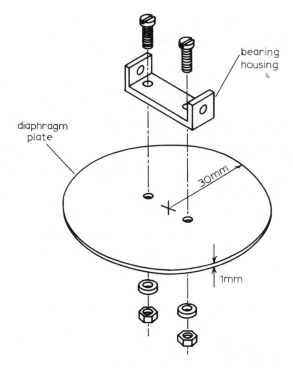

Fig. 8.14 Diaphragm plate and bearing housing assembly (original design).

Fig. 8.13) is used for automatic assembly processes. The procedure in analyzing the design of an assembly is to consider each individual handling and assembly operation separately in the order of the chosen assembly sequence. It is therefore assumed that the designer has been able to ascertain a suitable sequence for the various assembly operations. For each individual handling and assembly operation, the part involved is classified by a three-digit number selected according to the automatic handling code and the assembly process is coded according to the two-digit automatic assembly code. The codes thus obtained are inserted into successive lines on a worksheet in the order in which the operations are performed, together with certain supplementary information requested on the worksheet.

Figure 8.14 shows the original design of a diaphragm plate and bearing housing assembly and Fig. 8.15 shows how the worksheet would be completed. The following are the steps taken in analyzing the design of this product and completing the worksheet.

ID of new sub-assembly	Number of repeated operations	Part ID	Handling process code	ID(s) of part(s) or fixture to which part is added	ID(s) of other part(s) located or secured	Assembly process code	Remarks
0	1	2	3	4	5	6	
	02	60	001	90		00	place nuts in fixture
	02	61	001	60	60	00	place washers
		00	008	90	61	02	place diaphragm plate
		01	834	00		08	place bearing housing
02	02	62	210	00	01	69	insert screws

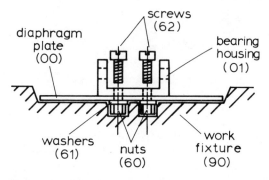

Fig. 8.15 Worksheet for diaphragm plate and bearing housing assembly (original design).

The first step is to number the work fixture and the various parts of the assembly (some of which may be subassemblies). The successive lines on the worksheet can now be completed, starting with the placement of the nuts on the work fixture and ending with the completion of the assembly.

For the first operation, the part identification number (ID), which is 60 for the nuts, is entered in column 2. The handling process code (001) for the nuts is entered in column 3, and since the nut is added to the work fixture, the ID for the work fixture (90) is entered in column 4. The assembly process code (00 for the nuts) is entered in column 6. Since the operation is to be repeated, the number 02 is entered in column 1. It will be noted that since a new subassembly was not completed by the operation and no other parts were secured by the operation, columns 0 and 5 were left blank.

The second line of the worksheet is completed in a similar way for the operation of placing the washers on the nuts. The third line of the worksheet is completed for the placing of the diaphragm plate on the work carrier, and so on. Also, since the insertion of the screws results in a completed assembly or subassembly, an identification number for the new subassembly is entered in column 0. This identification number can be used in subsequent analyses where the plate and bearing housing assembly might be treated as a single part. The operation of inserting the screws also secured the bearing housing, and therefore the identification number for the bearing housing is entered in column 5. The remaining columns are completed as before. It may be noted that the second digit of 9 for the assembly process code indicates a time-consuming screwing operation and a situation where the part is not easy to align and position during insertion.

It is suggested that designers, as they complete the worksheet, identify those digits that appear to indicate areas where redesign might be considered (these digits are circled in Fig. 8.15). The basic idea would be to attempt to reduce the digits of the code numbers where a reduction in automation difficulty is implied. In the present example some of the redesign considerations would be as follows:

1. A reduction in the number of parts in a product or assembly should usually be the first objective of a designer wishing to reduce assembly costs. In the present example, the difficulty of automatic assembly would be reduced significantly if the housing could be made self-securing. Two alternative designs that do not require separate fasteners and would be preferred for automation are shown in Fig. 8.16.

2. Assuming that separate fasteners are necessary, consideration could now be given to alternative fastener designs. For example, rivets or other fasteners that require fewer fastening elements and present less difficulty in automatic assembly might be employed. If, however, because of disassembly considerations, it is necessary to employ screws, consideration could be given to the use of screw points that facilitate alignment and thread-starting. Such designs have been found to reduce automatic assembly problems considerably. Consideration could be given to the elimination of the nut, the use of a threaded insert pressed into the diaphragm plate, or perhaps, a thread tapped in the plate itself. Combining the nut and washer would also simplify the assembly problem.

3. The automatic handling code for the original design of the diaphragm plate indicates that a redesign of this part should be con-

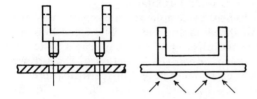

Integral rivet design

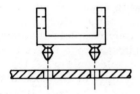

Snap fit design

Fig. 8.16 Possible redesigns without the use of separate fasteners.

Fig. 8.17 Redesign of diaphragm plate to reduce automatic handling difficulty.

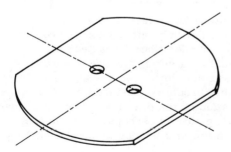

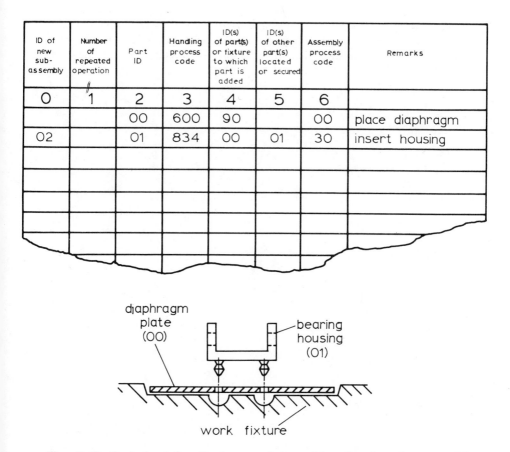

ID of new sub-assembly	Number of repeated operation	Part ID	Handling process code	ID(s) of part(s) or fixture to which part is added	ID(s) of other part(s) located or secured	Assembly process code	Remarks
O	1	2	3	4	5	6	
		OO	600	90		OO	place diaphragm
02		01	834	OO	01	30	insert housing

Fig. 8.18 Worksheet for diaphragm plate and bearing housing assembly (proposed design).

sidered. The third digit of the part's handling code is 8, indicating a part difficult to handle automatically. One possible redesign to reduce automatic handling difficulty is shown in Fig. 8.17. This redesigned plate has an automatic handling code of 600, indicating a part easy to handle automatically. In this example, assuming that separate fasteners are not really necessary for holding the housing on the plate, one of the possible redesigns and its associated worksheet are shown in Fig. 8.18. The code number spectrum for this design indicates a situation where automation could be easily applied.

It should be mentioned that, at the time of writing, improvements were being made to the procedures described here. In particular, techniques for obtaining a numerical rating for a design were being developed. However, it is hoped that the description above will serve to indicate the general philosophy behind the development of the design for assembly handbook.

Chapter 9

Design of Assembly Machines

The subject of this chapter brings together the results of work in all the various aspects of automatic assembly. Using the knowledge resulting from research, development, and experience, the designer of an assembly machine must produce a proposal that combines many requirements. Some of these requirements such as reliability and durability, are similar to those for any machine tool. However, certain requirements are applicable only to the assembly machine and are mainly a result of the variations in the quality of the component parts to be assembled. It can reasonably be assumed that an assembly machine can be designed which, if it is only fed with carefully inspected parts, will repeatedly perform the necessary assembly operations satisfactorily. Sometimes, unfortunately, the real problems of automatic assembly appear only when the machine is installed in the factory and the firm fills the feeders with its own standard parts containing the usual proportions of defective parts. The possible effects of feeding a defective part into an assembly machine are:

1. The mechanical workhead may be seriously damaged, resulting in several hours or even days of downtime.
2. The defective parts may jam in the feeder or workhead and result in machine or workhead downtime while the fault is cleared.
3. The part may pass through the feeder and workhead and spoil the assembly, thus effectively causing downtime equal to one machine cycle and producing an assembly that must be repaired.

Often, large samples of the parts to be assembled are available when the assembly machine is designed and clearly the designer should take into account, at the design stage, the quality levels of the parts

and the possible difficulties resulting from them. Even the choice
of basic transfer system can significantly affect the degree of difficulty
caused by defective parts, and therefore most of the important decisions
will be made by the designer before detailed design is considered.
Basically, the object of the design should be to obtain minimum downtime on the machine resulting from a defective part being placed in
the corresponding feeding device. This was illustrated in Chapter 7,
which dealt with the economics of multistation indexing assembly
machines. It was shown that the minimum cost of assembly, $C_{t(min)}$,
obtained when parts of optimum quality level are used, is given by

$$C_{t(min)} = Mt + 2n(MBT)^{1/2} + \sum_{i=1}^{n} A_i \qquad (9.1)$$

where M is the cost of operating a machine per unit time if only acceptable assemblies are produced, t the cycle time of the machine, n the
number of parts to be assembled, B the factor indicating increased
cost of parts with higher quality level, T the machine downtime due to
one defective part, and A_i the basic cost of one part. Thus any reductions in T, the time taken to clear a fault caused by a defective
part, will reduce the assembly costs by reducing the total machine
downtime. The first section of this chapter discusses the various design factors that can help to ensure minimum downtime due to defective
parts, and the second section describes in detail the design feasibility
study of a typical assembly problem.

9.1 Design Factors to Reduce Machine Downtime Due to Defective Parts

The first object in designing feeders and mechanisms for use in automatic assembly is to ensure that the presence of a defective part will
not result in damage to the machine. This possibility does not generally
exist where the part is moving under the action of its own weight
(that is, sliding down a chute) or being transported on a vibrating
conveyor. However, if the part is being moved or placed in a positive
way, it is necessary to arrange that the desired motion is provided
by an elastic system. In this case, if a defective part becomes jammed,
motion can be taken up in the spring members. For example, if a
plunger is to position a part in an assembly, it would be inadvisable
to drive the plunger directly by a cam. It would be better to provide
a spring to give the necessary force to drive the plunger and use
the cam to withdraw the plunger. With this arrangement a jammed
part could not damage the mechanism.

The next object in design should be to ensure that a jammed part can be removed quickly from the machine. This can be facilitated by several means, some of which are as follows:

1. All feeders, chutes, and mechanisms should be readily accessible. External covers and shields should be avoided wherever possible.
2. Enclosed chutes, feeders, and mechanisms should not be employed. Clearly, one of the cheapest forms of chute is a tube down which the parts can slide freely to the workhead. However, a jam occurring in a closed tube is difficult to clear. Although probably more expensive to provide, open rails are preferable in this case so that the fault can be detected and cleared quickly.
3. An immediate indication of the location of a fault is desirable. This may be achieved by arranging that a warning light is switched on and a buzzer operated when any operation fails. If the warning light is positioned at the particular workhead, the machine operators will be able to locate the fault quickly.

It is necessary for the machine designer to decide whether to arrange that the machine is stopped in the event of a fault or whether to arrange that the spoiled assembly continue through the machine. In Chapter 7 we saw that in typical circumstances, it is preferable to keep the machine running. However, it would be clearly undesirable to attempt further operations on the spoiled assembly. For this purpose the *memory pin* system can be employed, where each work carrier is fitted with a pin which, in the event of a failure, is displaced by a lever fitted to the workhead. Each workhead is also provided with a feeler which senses the position of the pin prior to carrying out the operation. If the pin is displaced, the operation is not carried out. A microprocessor can accomplish this same effect with appropriate software and electronic control.

The difficulty with the memory pin system is that it is not possible for the workhead to detect immediately whether a particular fault will be repetitive. One possibility is to arrange that initially, any type of fault will displace the memory pin but that when two or three faults have occurred in succession, the machine is automatically stopped. Alternatively, it may be left to a machine operator to observe when a succession of faults occurs and for the operator to stop the machine.

The discussion above dealt with methods of reducing the downtime on a machine caused by defective parts. Clearly, the ideal situation would be when the defective parts are detected and rejected in the feeding devices. Although it is generally not possible to perform complete inspection during the feeding of parts, it is sometimes possible to eliminate a considerable proportion of the defective ones. Figure 9.1 shows an example given by Ward where unsawn bifurcated rivets

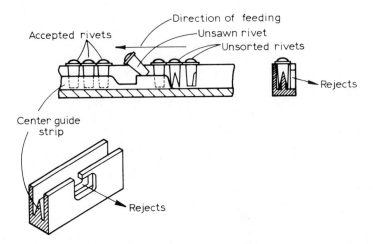

Fig. 9.1 System for inspecting bifurcated rivets. (From K. A. Ward, "Fastening Methods in Mechanized Assembly," paper presented at the Conference on Mechanized Assembly, July 1966, Royal College of Advanced Technology, Salford, England.)

Fig. 9.2 System for rejecting foreign matter in vibratory bowl feeder. (From K. A. Ward, "Fastening Methods in Mechanized Assembly," paper presented at the Conference on Mechanized Assembly, July 1966, Royal College of Advanced Technology, Salford, England.)

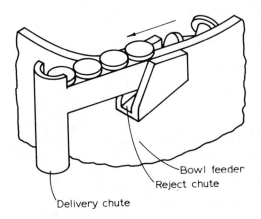

are detected and rejected in a bowl feeder. In this case the device was incorporated to prevent unsawn rivets being fed to the riveting head, where they would damage the mechanism during the operation. Sometimes, small pieces of swarf and other foreign bodies find their way into a bowl feeder, and a simple way of rejecting these is illustrated in Fig. 9.2. Quite sophisticated inspection arrangements can be built into a bowl feeder and it has been found economical in some circumstances to develop such a device to inspect the parts before they are placed in the assembly machine feeders. In this case the downtime occurs on the inspection device instead of on the assembly machine. This system is unlikely to be economical unless the inspection device is able to supply parts for several workheads or several assembly machines.

9.2 Feasibility Study

The decision to build or purchase an automatic assembly machine is generally based on the results of a feasibility study. The object of this study is to predict the performance and economics of the proposed machine. In automatic assembly these predictions are likely to be subject to greater errors than with most other types of production equipment, mainly because the machine is probably one of a kind and its performance will be very dependent on the qualities of the parts to be assembled. Also, similar machines will not generally be available for study. Nevertheless, a feasibility study must be made and all the knowledge and experience acquired in the past from automatic assembly projects must be applied to the problem in order to give predictions that are as accurate as possible.

Certain information is clearly required before a study can be made. For example, maximum and minimum production rates during the probable life of the machine must be known. The range of variations in these figures is very important because a single assembly machine is very inflexible. The operators required on the machine must all be present when the machine is working, or if the machine is stopped due to a falloff in demand for the product, they must be employed elsewhere. Thus, automatic assembly machines are generally suitable only when the volume of production is known to be steady. Further, they can usually only be applied profitably when the volume is high. Tipping* suggests that where the volume is 500,000 assemblies per

*W. V. Tipping, "Design of Mechanized Assembly Lines," paper presented at the Conference on Mechanized Assembly, July 1966, Royal College of Advanced Technology, Salford, England.

year or more, the application of automatic assembly has an excellent chance of success. Apart from this high volume requirement, the labor costs of the existing assembly process must also be high if automatic assembly is to be successful.

Clearly, much more information will be required and many other factors will combine to determine the final answer. Some of these are discussed in greater detail below.

9.2.1 Precedence Diagrams

It is always useful when studying the assembly of a product to draw a diagram which shows clearly and simply the various ways in which the process may be carried out. In most assemblies there are alternatives in the order in which some of the parts may be assembled. There are also likely to be some parts where no flexibility in order is allowed. For example, in the three-pin power plug shown in Fig. 9.3, the pins may be placed in position in any order, but the fuse can only be inserted after the fuse clip and the live pin are in position. Further, the cover can be placed in position and secured only after all the remaining parts have been assembled into the base. The precedence diagram is designed to show all these possibilities and its use has been

Fig. 9.3 Assembly of three-pin power plug.

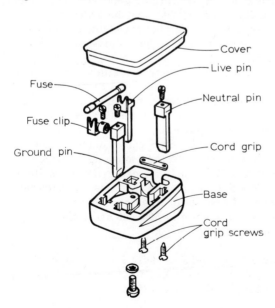

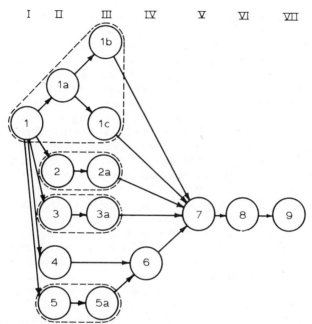

I II III IV V VI VII

1. Load base onto work carrier		4. Live pin	
1a. Cord grip		5. Fuse clip	
1b. Cord grip screw		5a. Fuse clip screw	
1c. Cord grip screw		6. Fuse	
2. Ground pin		7. Cover	
2a. Ground pin screw		8. Cover screw	
3. Neutral pin		9. Remove complete assembly	
3a. Neutral pin screw			

Fig. 9.4 Precedence diagram for complete assembly of power plug.

described in detail by Prenting and Battaglin.* A precedence diagram for the assembly of the power plug assuming that no subassemblies are involved is shown in Fig. 9.4, where it can be seen that each individual operation has been assigned a number and is represented by an appropriate circle with the number inscribed. The circles are connected by arrows showing the precedence relations.

In drawing the precedence diagram, all the operations that can be carried out first are placed in column I. Usually, only one operation

---

*T. O. Prenting and R. M. Battaglin, "The Precedence Diagram: A Tool for Analysis in Assembly Line Balancing," *J. Ind. Eng.*, vol. 15, no. 4., July-Aug. 1964, p. 208.

appears in this column: the placing of the base part on the work carrier. Operations that can only be performed when at least one of the operations in column I has been performed are placed in column II. Lines are then drawn from each operation in column II to the preceding operations in column I. In the example in Fig. 9.4, none of the column II operations can be performed until the base of the power plug has been placed on the work carrier, and therefore lines are drawn connecting operations 1a, 2, 3, 4, and 5 to operation 1. Third-stage operations are then placed in column III with appropriate connecting lines, and so on until the diagram is complete. Thus, following all the lines from a given operation to the left indicates all the operations that must be completed before the operation under consideration can be performed.

In the assembly of the power plug there are 15 operations and it will probably be impracticable to carry out all these on a single machine. For example, it would be difficult to assemble the cord grip and the two cord grip screws while the plug base is held in a work carrier because these parts enter the base from different directions. It is probably better, therefore, to treat these parts as a subassembly, and this is indicated in Fig. 9.4 by the dashed line enclosing the necessary operation. In a similar way the neutral and earth pins and the fuse clip together with their respective screws can also be treated as subassemblies. These groups of operations are all indicated by the dashed lines in the figure.

One of the objectives in designing an automatic assembly machine should be to include as few operations as possible on the line in order to keep machine downtime to a minimum. It is desirable, therefore, to break the product down into the smallest number of subassemblies and carry out individual studies of the subassemblies. If these can be mechanically assembled, separate machines may be used. These machines can then be arranged to feed the main assembly machine at the appropriate points.

Figure 9.5 shows the precedence diagram for the subassemblies of the power plug. It can be seen that no flexibility exists in the ordering of operations 1, 7, 8, and 9. Operations 2, 3, 4, 5, and 6, however, can be carried out in any order between operations 1 and 7 except that 6 cannot be performed until both 4 and 5 are completed. Considering the group of operations 4, 5, and 6 first, there are two ways in which these can be performed; either 4, 5, 6 or 5, 4, 6. Operation 3 could be performed at any stage in this order, giving $4 \times 2 = 8$ possibilities. Finally, operation 2 could be performed at any stage in the ordering of operations 3, 4, 5, and 6, giving a total of $5 \times 8 = 40$ possibilities. Thus, the precedence diagram shown in Fig. 9.5 represents 40 possible orderings of the various assembly operations.

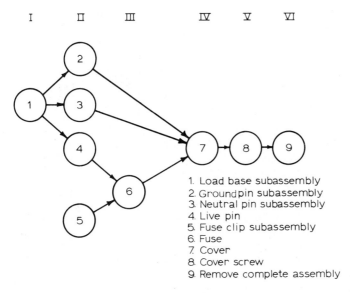

I   Ⅱ   Ⅲ   Ⅳ   Ⅴ   Ⅵ

1. Load base subassembly
2. Ground pin subassembly
3. Neutral pin subassembly
4. Live pin
5. Fuse clip subassembly
6. Fuse
7. Cover
8. Cover screw
9. Remove complete assembly

Fig. 9.5 Precedence diagram for assembly of power plug subassemblies.

In a fixed-cycle or indexing assembly machine, the order of assembly may not be a very important factor, but the chosen order of assembly in a free-transfer machine can have an effect on machine performance. However, before these questions can be considered it will be necessary to estimate the quality levels of the parts to be assembled.

9.2.2 Quality Levels of Parts

If assembly of a completely new product is to be contemplated, the estimation of the quality levels of the parts may be extremely difficult, if not impossible. However, a large proportion of assembly machine feasibility studies are concerned with existing products, and in these cases experiments can be performed to determine the quality levels of the various parts. It should always be remembered in such a study that defective parts do not generally create great difficulties when assembly is by hand. The operator can often quickly detect and reject a defective part and in many cases, when the "defective part" is simply a foreign body such as a piece of swarf or a cigarette end, the operator will not even attempt to pick it up but will simply leave it in the parts container to be discarded later. This means that a study of quality level must be conducted at the existing assembly stations where the numbers of discarded parts and foreign bodies can be recorded. A further danger is that many engineers responsible for assembly proc-

Table 9.1 Quality Levels of Power Plug Parts (Hypothetical)

Parts	Fault	Number of faults in assembling 10,000 plugs	Percentage faults
Base subassembly	Chipped	10	0.10
	Earth pin will not assemble	170	1.70
	Live pin will not assemble	20	0.20
	Neutral pin will not assemble	30	0.30
Earth pin subassembly	No screw	41	0.41
Neutral pin subassembly	No screw	59	0.59
Live pin	Fuse will not assemble	123	1.23
	Fuse assembles unsatisfactorily	21	0.21
Fuse clip subassembly	Fuse will not assemble	115	1.15
	Fuse assembles un-satisfactorily	17	0.17
Fuse	Damaged	18	0.18
Cover	Chipped	10	0.10
	Cover screw hole blocked	200	2.00
Cover screw	No thread or slot	20	0.20

esses assume that 100% visual inspection results in 100% acceptable parts. The assumption that an operator inspecting every part that is to be assembled will detect every defective part is clearly not valid.

The best procedure in estimating quality levels is for the investigator to sit with the assembly operators for a substantial period of time and note every defective part or foreign body that is discarded. Obviously, it will be inadvisable to assume that the quality levels recorded cannot be improved on, but it will be necessary to estimate the cost of these improvements and allow for this extra cost in the feasibility study.

Having noted the number of defective parts in a given batch, it will then be possible for the investigator to divide these into two categories: (1) those parts that cannot be assembled, for example screws with no thread or slot; and (2) those parts that can be assembled but are normally rejected by the operator, for example discolored or chipped parts.

The number of parts falling within the first category will allow estimates to be made of the assembly machine downtime, and those falling within the second category will allow estimates to be made of the number of unacceptable or defective assemblies produced by the machine.

A hypothetical set of figures for the power plug shown in Fig. 9.3 are presented in Table 9.1. It is important to remember that no assessment can be made of the most suitable type of assembly machine or of the number of operations that can economically be performed mechanically until the individual quality levels of the various parts have been investigated.

9.2.3 Parts Feeding and Assembly

An estimate must now be made of the degree of difficulty with which the individual parts can be automatically fed and assembled. It should be noted here that for each operation, four possibilities exist:

1. Automatic feeding and assembly
2. Manual feeding and automatic assembly
3. Automatic feeding and manual assembly
4. Manual feeding and assembly

At this stage in the feasibility study it may be necessary to resort to experiment. In considering the feeding of parts, all but the simplest shapes will probably require a vibratory bowl feeder, and simple experiments can normally be performed to test various ideas for orienting and feeding. Estimates can then be made of the various feed rates possible. For a given bowl feeder the maximum feed rate obtainable is proportional to the reciprocal of the length of the part, assuming that all parts arrive at the bowl outlet, end to end. Thus, with large parts that have many possible orientations, only one of which will be required, the feed rate of oriented parts can be very low.

For example, both the base and top of the power plug shown in Fig. 9.3 are approximately 50 mm long and can be fed up a bowl feeder track in any of eight possible orientations. Thus, if suitable devices were fitted to the track of the bowl to reject seven of the eight orientations and the bowl has a maximum useful conveying velocity of 50 mm/s, the maximum possible feed rate of oriented parts will be given by Eq. (5.32):

$$F_{max} = \frac{vE}{A}$$

(9.2)

where v is the mean conveying velocity, A the length of part, and
E the modified efficiency of the orienting system [see Eq. (5.33)]
and is roughly given by the number of acceptable orientations of the
part divided by the total number of stable orientations of the part.
Thus, in the present example,

$$F_{max} \simeq \frac{50(1/8)}{50} = 0.125 \text{ part/s } (7.5 \text{ parts/min})$$

In this approximation it has been assumed that there is an equal
possibility of the part being conveyed in each of its eight possible
modes of orientation. Clearly, this is not and will generally not be
the case, and if the orienting devices are designed to reject the orientations of least probability, the maximum feed rate will be higher. Further, if active orienting devices are fitted which can correct some of
the undesired orientations, the yield of correctly oriented parts can
be increased. The possibility also exists that some simple change in
the design of the parts will either reduce the number of orientations
or increase the probability of the desired orientation.

Bearing all these points in mind, the investigator will make a decision as to whether automatic feeding of the particular part is feasible.
In the example of the power plug it is possible that the base subassemblies and cover could not be fed at the required rate and that the
fuse clip assembly, because of its complicated shape, could not be
handled economically by automatic means. The remaining parts and
subassemblies could probably all be fed and assembled automatically
with bowl feeders and placing mechanisms, with the exception of the
main holding screw. This could be fed and screwed from below with
a proprietary automatic screwdriver.

9.2.4 Machine Layout and Performance

Three main possibilities exist for the layout of this assembly machine:
(1) in-line indexing, (2) rotary indexing, and (3) in-line free transfer.

In-Line Indexing Machine. If it is assumed that the base, top,
and fuse clip are to be assembled manually on a straight in-line machine,
at least two operators will be required on the assembly machine. The
first, positioned at the beginning of the line, could place the base
subassembly on the work carrier and place the fuse clip assembly in
the base (operations 1 and 5 of Fig. 9.5, respectively). The second
operator could assemble the cover and remove the complete plug assembly from the end of the line (operations 7 and 9 of Fig. 9.5, respectively).

It is generally necessary on an assembly machine to include some
inspection stations. In the present example, it is clear that after the

plug cover has been assembled, there will be no simple means of inspecting for the presence of the fuse clip, fuse, and the three small screws in the neutral and earth pins and the fuse clip. Thus, it will be necessary to include an inspection head on the machine immediately before operation 7 (the assembly of the cover), which will check for the presence of all these parts.

In the present example it will also be necessary to decide whether the inspection head should be designed to stop the machine in the event of a fault or to prevent further operations being performed on the assembly. In the hypothetical studies presented for each design of machine it will be assumed that the memory system is incorporated where the inspection head will be designed to activate the memory system rather than stop the machine.

The general layout of a typical in-line indexing machine is shown in Fig. 9.6. It will be noted that operations 4 and 6 have been arranged

Fig. 9.6 Station layout of in-line indexing machine for assembling power plugs.

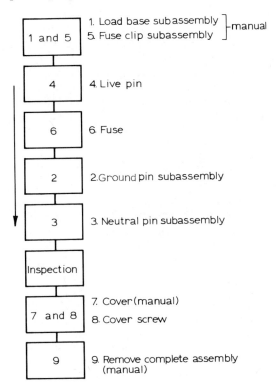

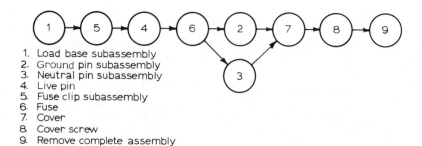

1. Load base subassembly
2. Ground pin subassembly
3. Neutral pin subassembly
4. Live pin
5. Fuse clip subassembly
6. Fuse
7. Cover
8. Cover screw
9. Remove complete assembly

Fig. 9.7 Final precedence diagram for assembly of power plug.

directly after the first (operator) station. This is to minimize the possibility of the fuse clip becoming displaced during the machine index. When the fuse is in position, the fuse clip is then positively retained. These desirable features provide further restrictions in the order of assembly, and the precedence diagram is modified to that shown in Fig. 9.7.

The downtime on an indexing machine is given by the sum of the downtimes on the individual heads due to the feeding of defective parts plus the effective downtime due to the production of unacceptable assemblies.

If for each machine station, x is the effective proportion of defective to acceptable parts, then mx is the average proportion of defectives that will cause a machine stoppage and $(1 - m)x$ is the effective average proportion of defectives that will spoil the assembly but not stop the machine. The downtime due to machine stoppages and the final production rate are found as follows.

In producing N assemblies the number of machine stoppages is $N \Sigma mx$, where Σmx is the sum of the individual values of mx for the automatic workheads.

If T is the average time to correct a fault and restart the machine, the downtime due to machine stoppages is $NT \Sigma mx$; if t is the machine cycle time, the proportion of downtime D will be given by

$$D = \frac{\Sigma mx}{t/T + \Sigma mx} \tag{9.3}$$

The figures in Table 9.1 are rearranged in Table 9.2 to give the effective quality levels for the various operations. From these figures it can be seen that the value of Σmx is 0.0678, and assuming that t = 3 s (the time taken to place the base and assemble the fuse clip manually) and T = 15 s (a typical figure in practice), then

Table 9.2 Effective Quality Levels in Assembly of Power Plug

Operation	Automatic station on free-transfer machine	Effective quality level, x	Ratio of defectives causing machine stoppages, m	mx	(1 − m)x
1. Assemble base subassembly onto work carrier	—	0.001	0	0	0.001
2. Assemble earth pin sub-assembly into base	4	0.017	1.0	0.017	0
3. Assemble neutral pin sub-assembly into base	3	0.003	1.0	0.003	0
4. Assemble live pin into base	1	0.002	1.0	0.002	0
5. Assemble fuse clip sub-assembly into base	—	0	0	0	0
6. Assemble fuse into live pin and fuse clip	2	0.0294	0.813	0.0238	0.0056
7. Assemble cover	5	0.001	0	0	0.001
8. Assemble cover screw	5	0.022	1.0	0.022	0
9. Remove complete assembly	—	0	—	—	—
10. Inspection	—	0.01	0	0	0.01
				0.678	0.0176

$$D = \frac{0.0678}{0.2 + 0.0678} = 0.253$$

During the time the machine is operating, some of the assemblies produced will contain defective parts which did not stop the machine, and assuming that no assembly contains more than one such defective part, the production rate of acceptable assemblies P_a will be given by

$$P_a = \frac{[1 - \Sigma (1 - m)x](1 - D)}{t} \tag{9.4}$$

From Table 9.2, $\Sigma(1 - m)x = 0.0176$, and therefore from Eq. (9.4),

$$P_a = \frac{(0.9824)0.747}{3} = 0.24 \text{ assembly/s } (14.7 \text{ assemblies/min})$$

Rotary Indexing Machine. The basic layout of a suitable rotary indexing machine is shown in Fig. 9.8. It has been assumed in this case that operations 1, 5, 7, and 9 would all be carried out by a single operator with an increased machine time cycle of t = 3.2 s. It is com-

Fig. 9.8 Station layout of rotary indexing machine for assembling power plugs.

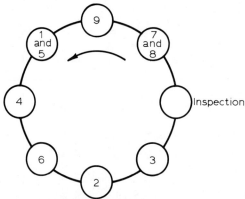

1. Load base subassembly (manual)
2. Ground pin subassembly
3. Neutral pin subassembly
4. Live pin
5. Fuse clip subassembly (manual)
6. Fuse
7. Cover (manual)
8. Cover screw
9. Remove complete assembly (manual)

monly accepted that six is the practical maximum number of stations
on a rotary indexing machine and it can be seen that in the present
example eight stations are required. For this reason it may be decided
that the machine will be too crowded with equipment to allow sufficient
access to the various points where parts may become jammed. The
effect of this will be to increase the value of T and it will be assumed
in the following analysis that T is increased to 16 s, thus maintaining
the ratio $t/T = 0.2$ as with the in-line indexing machine.

The machine downtime due to stoppages will remain at 0.253 as in
the previous example, but since $t = 3.2$ s, the production rate of
acceptable assemblies given by Eq. (9.4) will change to

$$P_a = \frac{(0.9824)0.748}{3.2} = 0.23 \text{ assembly/s (13.8 assemblies/min)}$$

In-Line Free-Transfer Machine. The layout of a free-transfer
machine suitable for assembling the power plug is shown in Fig. 9.9.
Progress has recently been made in reducing the high cost of free-
transfer machines by a unit construction principle. In this case it is
necessary to standardize on the size of buffer stock, and it will be
assumed in the following that each unit of the machine is capable of
accommodating five work carriers, one of which is situated below the
workhead, the remaining four comprising the buffer stock for the
workhead.

This type of machine was analyzed in Chapter 7, and the equations
presented in Table 9.3 apply to the present example. In the equations,
the effect of the first and last operations has been ignored, since
these operations of placing the base and fuse clip into the work carrier
and removing the completed assembly will not generally cause downtime on the machine. Operation 2 has been positioned nearer the end
of the machine than operation 3 since operation 3 deals with parts of
a higher quality level. This follows from the work in Chapter 7, which
indicated that to obtain minimum downtime, operations involving parts
with poor effective quality levels should be positioned as close to either
end of the machine as possible. Two operators will be required on
the machine under consideration and the original times of $t = 3$ s and
$T = 15$ s will be applicable.

Substitution in the equations in Table 9.3 of the appropriate values
of m_1x_1 to m_5x_5 inclusive from Table 9.2, writing $b = 4$ and setting
$d_1 = d_2 = d_3 = d_4 = d_5$, leads to the following values of a_1 to a_4:

$$a_1 = 3.88, \quad a_2 = 2.48, \quad a_3 = 2.3, \quad a_4 = 1.3$$

It is now found that, with these values, the last term in the first
two equations, the first term in the third and fourth equations, and

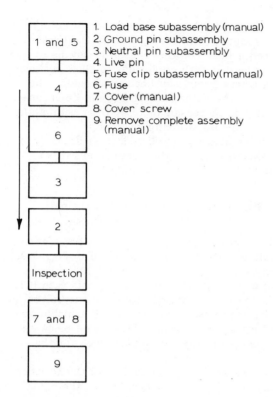

Fig. 9.9 Station layout of free-transfer assembly machine for assembling power plugs.

the first two terms in the last equation must now be omitted because they become negative. Resolving the equations now gives

$$a_1 = 3.82, \quad a_2 = 2.51, \quad a_3 = 2.37, \quad a_4 = 2.04$$

Substitution of these results in the last equation with the first two terms omitted gives

$$\frac{d}{Nt} = 0.59 \times 0.3 + 2.96 \times 1.7 + 5 \times 2.2 = 16.23$$

Thus,

$$D = \frac{0.1623}{1 + 0.1623} = 0.139$$

Table 9.3 General Equations for the Downtime d at Each Station of the Five Station Assembly Machine When Producing N Assemblies

Work-head	Downtime due to First workhead	Second workhead	Third workhead	Fourth workhead	Fifth workhead
First: $\frac{100d_1}{N} =$	Tm_1x_1	$+ [T-(b-a_1)t]m_2x_2$	$+ [T-(2b-a_1-a_2)t]m_3x_3$	$+ [T-(3b-a_1-a_2-a_3)t]m_4x_4$	$+ [T-(4b-a_1-a_2-a_3-a_4)t]m_5x_5$
Second: $\frac{100d_2}{N} =$	$[T-a_1t]m_1x_1$	$+ Tm_2x_2$	$+ [T-(b-a_2)t]m_3x_3$	$+ [T-(2b-a_2-a_3)t]m_4x_4$	$+ [T-(3b-a_2-a_3-a_4)t]m_5x_5$
Third: $\frac{100d_3}{N} =$	$[T-(a_1+a_2)t]m_1x_1$	$+ [T-a_2t]m_2x_2$	$+ Tm_3x_3$	$+ [T-(b-a_3)t]m_4x_4$	$+ [T-(2b-a_3-a_4)t]m_5x_5$
Fourth: $\frac{100d_4}{N} =$	$[T-(a_1+a_2+a_3)t]m_1x_1$	$+ [T-(a_2+a_3)t]m_2x_2$	$+ [T-a_3t]m_3x_3$	$+ Tm_4x_4$	$+ [T-(b-a_4)t]m_5x_5$
Fifth: $\frac{100d_5}{N} =$	$[T-(a_1+a_2+a_3+a_4)t]m_1x_1$	$+ [T-(a_2+a_3+a_4)t]m_2x_2$	$+ [T-(a_3+a_4)t]m_3x_3$	$+ [T-a_4t]m_4x_4$	$+ Tm_5x_5$

where D is the proportion of downtime due to machine stoppages. The production rate of acceptable assemblies will again be given by Eq. (9.4); thus,

$$P_a = \frac{(0.9824)(1 - 0.139)}{3} = 0.28 \text{ assembly/s } (16.9 \text{ assemblies/min})$$

9.2.5 Economics of the Various Machines

Having estimated the effective production rate for each type of machine, it is now possible to make comparisons of their economics.

The approach made here will be to assume that in the particular company concerned, an estimate can be made of the amount of capital that can be economically spent to replace each operator on one shift. Some companies put this figure at 30 k$, but clearly the economic amount can vary considerably depending on the operator overheads, availability of labor, and the amortization period for the proposed machine.

For each machine considered above, the number of acceptable assemblies produced in one day will be estimated. The extra number of operators required to produce this number of assemblies in the same period will then be calculated. It will be assumed that, on average, one operator can assemble 1000 plugs in one 8-hour shift, using the same subassemblies and parts as the proposed machine and provided with a jig and all other necessary aids costing 10 k$ per operator. These figures, although hypothetical, are considered realistic, and in a practical feasibility study, it is clearly important to obtain an accurate estimate of the operator costs and production rate with the manual process.

For each type of machine the various feeders, workheads, and inspection devices are identical, the only variation being in the transfer device employed. In all, five automatic workheads with feeders and one inspection head are required, and these will be assumed to cost a total of 75 k$ irrespective of the type of transfer machine.

Table 9.4 gives the assumed costs of the three transfer machines studied above, including control unit, work carriers, and where appropriate, work carrier return system, together with the effective production rate and the number of operators required. In each case one operator has been allowed for correcting machine faults and filling the feeders at regular intervals. The economics for each machine are also presented in Table 9.4.

With the purely hypothetical figures used in the example, the rotary index machine would appear to be the most economical. However, there was some doubt whether the eight stations required could be arranged round one rotary indexing machine, and if this were not possible, the in-line indexing machine would be preferred.

Table 9.4 Economics of Various Transfer Machines
in Power Plug Assembly

	In-line index	Rotary index	Free transfer
1. Cost of transfer device and work carriers (k$)	88	88	216
2. Cost of workheads, etc. (k$)	75	75	75
3. Total cost of machine [(1) + (2)] (k$)	163	163	291
4. Effective production rate (assemblies per minute)	14.7	13.8	16.9
5. Number of operators for three shifts	9	6	9
6. Number of assemblies produced in three 8-hour shifts	21,150	19,830	24,360
7. Number of operators required to produce (6) manually	21.2	19.8	24.4
8. Cost of equipment for manual operation [(7) × 10/3] (k$)	70.7	66.0	81.3
9. Effective cost of machine [(3) − (8)] (k$)	92.3	97.0	209.7
10. Number of operators saved [(7) − (5)]	12.2	13.8	15.4
11. Capital outlay per operator saved [(9) ÷ (10)] (k$)	7.57	7.03	13.62

single product has a market life of 3 years or more without significant fluctuations in demand, the manual fitting or adaption of parts is not required and the parts are of sufficiently high quality

investment in automation encouraged
SQ/W ≥ 3

few product styles Y ≤ 1.5		several product styles Y > 1.5	
few design changes $n_d \leq 0.5$	several design changes $n_d > 0.5$	few design changes $n_d \leq 0.5$	several design changes $n_d > 0.5$

			0	**1**	**2**	**3**
annual production volume per shift greater than 0.7 million assemblies $V_{as} > 0.7$	9 or more parts in the assembly $n \geq 9$	**0**	AF	AF	AP	AP
	less than 9 parts in the assembly $n < 9$	**1**	AI	AI	AI	AI
annual production volume per shift greater than 0.5 million assemblies $0.5 < V_{as} \leq 0.7$	25 or more parts in the assembly $n \geq 25$	**2**	AF	AP	AP	AP
	15 or more parts in the assembly $15 \leq n < 25$	**3**	AF AP	AP AF	AP	AP
	10 or more parts in the assembly $10 \leq n < 15$	**4**	AI	AI	AI	AP AI

Fig. 9.10 Portion of classification system for products and assemblies from G. Boothroyd, "Design for Assembly Handbook" (Department of Mechanical Engineering, University of Massachusetts, Amherst, Mass., 1979). AF is the free-transfer machine with special-purpose workheads and parts feeders, AI the indexing machine with special-purpose workheads and parts feeders, AP the free-transfer machine with programmable workheads and part magazines, $n_d = N_d/n$.

Clearly, in a practical situation other factors, such as incentive schemes and availability of operators, would need to be taken into account, but it is hoped that the example presented here will serve as a guide to the procedure for carrying out a detailed feasibility study for automatic assembly.

Included in the handbook "Design for Assembly" developed at the University of Massachusetts at Amherst is a classification system for products and assemblies, part of which is shown in Fig. 9.10. This classification system indicates which assembly system is likely to be the most economic for a given situation and is based on economic models similar to those described in Chapter 7. If this system is applied to the conditions used in the example presented in this chapter, that is, $n = 8$, $V_{as} = 1.8$ million assemblies per shift per year, $Y = 1$, and $n_d = 0$, an indexing machine is recommended.

Problems

1. A vibratory bowl feeder is fitted with springs that are inclined
   at an angle of 60 degrees to the horizontal. The upper portion
   of the bowl track is horizontal and the frequency of operation is
   60 Hz. The amplitude of vibration is adjusted such that the normal
   track acceleration has a maximum of 1.0g for the upper part of
   the track.
   a. If the coefficient of friction between the parts and the track
      is 0.5, will forward conveying be achieved by forward sliding
      only or by a combination of forward and backward sliding?
   b. What will be the maximum parallel track velocity, v_p(mm/s)?
   c. If the conveying efficiency η is 70%, what will be the mean
      conveying velocity v_m of the parts? [Note: $\eta = (v_m/v_p) \times 100$.]

2. A standard vibratory bowl feeder has three leaf springs inclined
   at 80 degrees to the horizontal. These springs are equally spaced
   around a circle of 225 mm radius and support a bowl that is 600
   mm in diameter.
   a. Determine the effective vibration angle for the horizontal
      upper part of the bowl track.
   b. If the peak-to-peak amplitude of vibration in the line of vibration
      at the bowl wall is 2.5 mm and the frequency is 60 Hz, deter-
      mine whether forward conveying will occur and whether this
      will be by both forward and backward sliding or by forward
      sliding only. (Assume that the coefficient of friction between
      part and track is 0.5.)
   c. If the feeding motion were 100% efficient, what would be the
      mean forward conveying velocity of the parts?

3. a. If a vibratory bowl feeder is driven at a frequency of 60 Hz
        and the vibration angle is 20 degrees, what peak-to-peak
        horizontal amplitude of vibration (mm) will give a mean con-
        veying velocity of 50 mm/s on a horizontal track?
    b. For the same conditions as those of part (a), what would be
        the minimum vertical clearance between a part and a wiper
        blade so that the wiper blade will never reject the parts?
    c. If the leaf springs on the feeder are mounted at a radial
        position r_2 of 100 mm and the track radius r_1 is 200 mm, what
        spring angle (measured from the horizontal) will give the
        vibration angle of 20 degrees?

4. A special decal on the side of a vibratory bowl feeder indicates
    that the horizontal peak-to-peak amplitude of vibration is 0.25 mm.
    The angle of the supporting springs is such that the vibration
    at the bowl wall is inclined at an angle of 20 degrees to the
    horizontal. The coefficient of friction between parts and track
    is 0.2, the frequency of vibration is 60 Hz, and the track is
    inclined at 5 degrees to the horizontal.
    Determine:
    a. The actual value of A_n/g_n for the inclined track
    b. The value of A_n/g_n for forward sliding to occur during the
        vibration cycle
    c. The value of A_n/g_n for backward sliding to occur
    d. Whether forward conveying will occur
    e. Whether hopping will occur

5. A vibratory bowl feeder has a spring angle of 70 degrees and is
    operated at a frequency of 60 Hz. The radius to the springs is
    150 mm and to the upper, horizontal part of the track is 200 mm.
    The amplitude of vibration is set so that the parts traveling around
    the upper track are on the verge of hopping.
    a. What is the horizontal (parallel) amplitude (a_p) of vibration
        at the bowl wall? Give the answer in μm.
    b. What is the minimum coefficient of friction between parts and
        track for forward conveying to occur?
    c. If the coefficient of friction is 0.3, what is the minimum value
        of the normal amplitude a_n (μm) for forward conveying to
        occur?

6. Derive an expression for the dimensionless maximum parallel
    distance jumped J/A_p^0 by a part during each cycle of vibratory
    feeding (J = parallel distance jumped relative to the track, A_p^0 =
    parallel track amplitude). Assume that for maximum distance
    jumped, $\omega t_2 = \omega t_1 + 2\pi$, where ω is the frequency in rad/s, t_1 the

time when the part leaves the track, and t_2 the time when the
part lands on the track. Also assume that the coefficient of
friction between the part and the track is such that the part
never slides. If the efficiency of conveying is defined by
$\eta = v_m(100)/v_{pmax}$, where v_m is the mean conveying velocity
and v_{pmax} the maximum parallel track velocity), calculate the
maximum efficiency possible under the conditions stated above
and when the track angle is zero.

7. A vibratory bowl feeder operates at a frequency of 60 Hz. The
track angle is 5 degrees, the vibration angle is 40 degrees, and
the horizontal component of the amplitude of vibration of the track
is 10 mm. If the coefficient of friction between the parts and the
track is 0.2, will the parts be conveyed up the track?

8. Parts in the form of right circular cones having a height H and
a base radius R are to be fed and oriented using a vibratory bowl
feeder. The bowl track is designed so that a part will be fed
to the orienting devices either standing on its base or lying on
its side with the base leading or following. Two active orienting
devices are to be used. The first is a step of height h_S, which
will cause a part standing on its base to overturn onto its side
but will not affect the remaining orientations. The second device
consists of a portion of track with a vee cross section so that all
parts lying on their sides will fall into the vee with their bases
uppermost.
    Determine the theoretical limits for the dimensionless height
of the step (h_S/H) for the situation described, assuming that a
very low conveying velocity is to be employed. Assume that the
parts do not bounce or slide as they are fed over the step, that
no energy is lost as the parts impact with the track, and that
$H = 4R$.

9. A rivet feeder (external gate hopper) has a gate (at the lowest
point in the feeder body) through which the rivets pass, 1.4d in
width, where d is the diameter of the shank of the rivet. The
rivets are tumbled within a rotating inner sleeve having slots to
accept the rivet shanks which are just wide enough to accept one
rivet. The gap between the inner sleeve and the body of the
feeder is equal to 0.6d and the distance between slots is 4.0d.
    a. Derive an expression, in terms of d, for the critical peripheral
        velocity of the inner sleeve to prevent jamming.
    b. Estimate the maximum feed rate in rivets per second if the
        efficiency with which rivets fall into the slots is 30% and the
        diameter of the rivet shanks d is 5 mm.

10.   A rivet feeder (external gate hopper) has a gate, through which
      the shanks of the rivets pass, 1.2d in width, where d is the
      diameter of the shank of the rivet. The rivets are tumbled within
      a rotating inner sleeve having slots to accept the rivet shanks
      which are just wide enough to accept one rivet. The gap between
      the inner sleeve and the body of the feeder is equal to 0.5d and
      the distance between slots is 2.5d.

      Estimate the maximum feed rate in rivets per second if the
      efficiency with which rivets fall into the slots is 30% and the
      diameter of the rivet shanks d is 5 mm. (Note: The gate is at
      the bottom of the feeder body.)

11.   A centerboard hopper has a blade length of 262.5 mm and is
      designed to feed cylindrical parts end to end. The center of
      rotation (which is in line with the track) is 250 mm from the lower
      end of the blade track. The inclination of the track when the
      blade is in its highest position is 45 degrees and the coefficient
      of sliding friction between the parts and the track is 0.3. The
      blade is driven by a cam drive such that the raising of the blade
      is carried out by a period of uniform acceleration followed by an
      equal period of deceleration of the same magnitude. The blade
      then remains stationary to allow parts to slide into the delivery
      chute and is finally lowered in the same manner as it was raised.

      If the efficiency with which 1-in. (25.4-mm)-long parts are
      selected by the blade track is 50%, calculate the maximum feed
      rate possible with these parts (parts per second) and the corre-
      sponding cycle time in seconds.

12.   A centrifugal hopper has a diameter of 0.5 m and is designed to
      feed steel cylindrical parts 20 mm long and 5 mm in diameter.
      The wall of the hopper is steel, and to increase the feed rate,
      the base is coated with rubber.

      If the coefficient of friction for steel/steel is 0.2 and for
      steel/rubber is 0.5, estimate:
      (a) The maximum possible feed rate for the hopper
      (b) The rotational frequency (rev/s) that will give this feed rate

13.   A centrifugal hopper has a diameter of 0.3 m and is designed to
      feed steel cylindrical parts 10 mm long and 4 mm diameter. The
      wall of the hopper is steel, and to increase the feed rate, the
      base is coated with rubber.

      If the coefficient of friction for steel/steel is 0.2 and for
      steel/rubber is 0.5, estimate:
      (a) The maximum possible feed rate for the hopper (parts/s)
      (b) The minimum rotational frequency (rev/s) that will give this
          feed rate

14. A vee-shaped orienting device for a vibratory bowl feeder is to be designed to orient truncated cone-shaped parts having a ratio of top diameter to base diameter of 0.8. Good rejection characteristics can be obtained when the half angle of the cutout is 45 degrees. Determine the maximum and minimum distances from the bowl wall to the apex of the cutout (expressed as a ratio of the part base radius) for which all the parts on their tops will be rejected and all the parts on their bases will be accepted. Assume that the parts never leave the track during feeding.

15. A vibratory bowl feeder is to feed cylindrical cup-shaped parts. The track is to be designed such that only four feeding orientations of the part are possible:

    a: part feeding on its base (27%)
    b_1: part feeding on its side, base leading (35%)
    b_2: part feeding on its side, top leading (35%)
    c: part feeding on its top (3%)

    The part has a length-to-diameter ratio of 1.13 and the orienting system comprises three devices:

    1. A step that does not affect 49% of the parts in orientation a and which reorients 100% of b_1, 28% of b_2, and 80% of c to orientation a
    2. A scallop cutout that rejects all the parts in orientation c
    3. A sloped track and ledge that rejects all parts in orientations b_1 and b_2 .

    If the input rate of parts to the orienting system is 2.5 per second, what will be the output rate of oriented parts from the system?

16. A part that is to be fed and oriented in a vibratory bowl feeder has five orientations (a, b, c, d, e) on the bowl track. The orienting system consists of three devices. The first is a wiper blade that rejects all the parts in orientation e; the second is a step that reorients 20% of those in orientation a to orientation b, 10% of those in b to orientation c, 50% of those in c to orientation a, and 80% of those in d to orientation a. The final device rejects all parts except those in orientation a. Write the matrices for each device and thereby calculate the efficiency of the system if the initial distribution of orientations is as follows:

    a: 0.2, b: 0.25, c: 0.4, d: 0.1, e: 0.05

What would the efficiency of the system have been if no step had been included?

17. Rectangular prisms are to be fed and oriented in a vibratory bowl feeder. The two orienting devices employed are a wiper blade and a narrowed track. The probabilities of the important orientations and the dimensions of the prisms are shown in Fig. P.17.
    a. Construct matrices for the wiper blade and narrowed track and hence find the system matrix.
    b. By multiplying the system matrix and the orientation matrix, find the efficiency of the orienting system.
    c. Estimate the feed rate of oriented parts (parts per second) if the conveying velocity is 100 mm/s and if the parts are contacting each other as they enter the system.

18. The part shown in Fig. P.18 is to be oriented and delivered using a vibratory bowl feeder. The orienting devices to be used are (1) a wiper blade, (2) a narrowed track, and (3) a scallop cutout.
    a. Estimate the probability of the delivery orientation.
    b. Estimate the feed rate of oriented parts if parts enter the orienting system at 1 per second on average.

19. A vibratory bowl feeder orienting system is designed to orient a part that has four orientations on the bowl track. The first device is a step whose performance can be represented by the following matrix:

$$
\begin{array}{c c c c c}
 & a & b & c & d \\
a & \begin{bmatrix} 0.49 & 0.51 & 0 & 0 \\ 1.0 & 0 & 0 & 0 \\ 0.28 & 0.02 & 0.3 & 0.4 \\ 0.80 & 0.1 & 0.1 & 0 \end{bmatrix}
\end{array}
$$

Fig. P.17

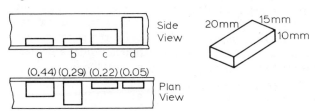

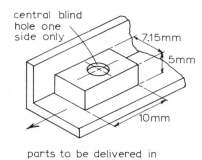

central blind
hole one
side only — 7.15mm

5mm

10mm

Fig. P.18

parts to be delivered in
orientation shown

The remaining devices are designed to reject orientations b, c, and d.

Determine the feed rate of oriented parts if the input rate to the system is 2.5 per second and if the input distribution of orientations is

a	b	c	d
[0.27	0.35	0.35	0.03]

Also estimate the effect on this performance if a further identical step were added to the beginning of the system.

20. Figure P.20 shows the orientations of a cup-shaped part that is to be fed by a vibratory bowl feeder. The orientation system consists of a step (for which the performance matrix is given) followed by a scallop that rejects all d's and a sloped track and ledge that rejects all b's and c's. The probabilities of the initial orientations a, b, c, and d are 0.45, 0.25, 0.25, and 0.05, respectively.
   a. What is the efficiency η of the orienting system?
   b. What would the efficiency be if the step were not used?

Fig. P.20

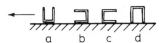

a   b   c   d

Matrix for step

	a	b	c	d
a	1	0	0	0
b	0.1	0.9	0	0
c	0.8	0	0.1	0.1
d	0.5	0	0.5	0

21. The rectangular prismatic part is to be fed and oriented in a vibratory bowl feeder as shown in Fig. P.21. Determine the efficiency (η percent) of this orienting system. Assume a hard track surface and assume that the coefficient of friction between the bowl wall and the part is 0.2.

22. The part shown in Fig. P.22 is to be fed and oriented by vibratory bowl feeder in the orientation depicted.

    A television system has been devised that will detect the orientation desired and trigger a mechanism that will reject all other orientations. What will the feed rate be? What is the part code? Assume that the coefficient of friction between the part and the bowl wall is 0.2. Assume that the conveying velocity of the parts on the bowl track is 20 mm/s and neglect the effects of the cutouts and hole on the probabilities of natural resting aspects and orientations. Also assume a hard surface on the bowl track.

23. A gravity feed track consists of a straight section of track inclined at an angle of 45 degrees followed by a section of constant radius 10 in. (254 mm) which is followed by a straight horizontal section

Fig. P.21

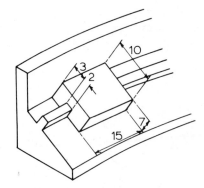

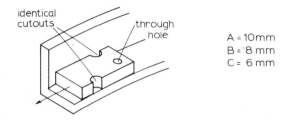

A = 10 mm
B = 8 mm
C = 6 mm

Fig. P.22

5 in. (127 mm) in length. The complete track is arranged in a vertical plane and is to be used in the feeding of parts 0.5 in. (12.7 mm) in length.

Determine the minimum vertical height (in inches or mm) of the top of the column of parts above the inlet to the workhead for feeding to occur if the coefficient of friction between the parts and the track is 0.4.

24. The cost of certain screws C is found to be given by

$$C = 0.1 + \frac{0.01}{x} \quad \$$$

where x is the percentage of faulty parts. A machine is to be built to assemble 20 such screws (one at each station) into an electrical terminal strip. An operator (cost $10/h) will be employed to load the terminal strips (cost $0.50) and an engineer (cost $20/h) to correct machine faults. The rate for the machine (excluding operator and engineer) is estimated to be $50/h.
   a. If all faulty parts cause a machine stoppage and it takes 20 s to correct each fault, find the *minimum* cost of each complete assembly on an indexing machine working on a 4-s cycle.
   b. If a robot could be used (cost $40/h) to feed, orient, and insert the terminal strips (quality level x = 1.5%) and the downtime for each faulty terminal strip was 20 s, what would the new minimum cost be?

25. An indexing assembly machine has 10 stations and the total cost of running the machine (including operators and overheads) is $0.02/s. The quality levels x of the parts at each station and the proportions of defective parts that cause a machine stoppage are as shown in Table P.25. Assume that the time T to correct a machine fault is 30 s on average and that operator assembly costs are

Table P.25

Station	1	2	3	4	5	6	7	8	9	10
x (%)	0.12	1.3	1.4	0.5	2.3	0.3	1.6	0.25	0.25	3.6
m	1.0	0.5	1.0	1.0	0	0.5	0.5	1.0	1.0	1.0

$0.25 per assembly. Assume also that the cost of disassembling unacceptable assemblies is $0.80 on average and that the machine has a memory pin system to detect those defective parts that do not stop the machine.

At what cycle time t (s) must the machine be operated so that assembly costs will be the same as those for manual assembly?

26. A 10-station indexing assembly machine is fully automated and each automatic workhead incorporates a memory pin system such that when a fault is detected, the machine continues to run but no further work is carried out on the assembly containing the fault. If the cycle time of the machine is 5 s and at each station the ratio of unacceptable to acceptable parts is x percent, calculate, approximately, the production rate of acceptable assemblies per minute when x = 1.5.

Given that the total running cost of the machine per minute is 1 dollar, it takes an average of 40 s to dismantle an unacceptable assembly, operator costs are $0.04/min, and the cost of each part used by the machine is given by A + B/x, where A = B = $0.004 per part, calculate the number of operators required to dismantle the incomplete assemblies and the total cost of the product. If the parts quality is now changed to the optimum value, determine this optimum and the new total cost of the product.

27. A five-station indexing assembly machine works on a cycle time of 3 s and the ratio of unacceptable to acceptable parts used at each station is 0.02. The average time to correct a fault at each station for a machine that stops is 20 s and the average time taken to dismantle each part of a faulty assembly on a machine that incorporates a memory pin system is 30 s. The total running cost of the machine is $0.40/min and the total cost of each operator used on the machine is W cents/min.
Determine:
a. The production rate and assembly cost per assembly for a machine that stops when a fault occurs.
b. The production rate and assembly cost per assembly for a machine that incorporates the memory system.

c. The value of W for which the cost per assembly is the same for both types of machine.

Assume that the machine that stops requires only one operator, and calculate the extra operators required on the machine that has a memory pin facility.

28. A five-station indexing assembly machine has a cycle time of 3 s. The quality levels of the parts supplied to the workheads is 1.1, 2.3, 0.6, 1.5, and 0.2%, respectively. The average time taken to correct a fault and restart the machine is 30 s. Assuming that every defective part will cause a machine fault, calculate:
    a. The percentage downtime on the machine
    b. The average production rate (assemblies per minute)
    If the cost of operating the machine (including overhead) were $20/h, what maximum amount could be spent over a 2-year period to eliminate the defective parts fed to station 2. (Assume 50 weeks working at 40 h/week.)

29. A five-station indexing assembly machine has a cycle time of 3 s. The quality level of the parts supplied to the workheads is 1.1, 2.3, 0.6, 1.5, and 0.2%, respectively. The average time taken to correct a fault and restart the machine is 30 s. Assuming that every defective part will cause a machine fault, estimate:
    a. The percentage downtime for the machine
    b. The average production rate (acceptable assemblies per minute)
    Repeat the calculations assuming that only half of the defective parts cause a machine fault, while the other half spoil the assembly.

30. A rotary indexing assembly machine has six stations. Station 1 involves the manual removal of the completed assembly and the manual loading of the base part of the assembly. The remaining stations are all automatic. These involve the assembly of three pins having quality levels of 1.4% defectives and a cover having a quality level of 0.9% defectives. At the last station, two holding-down screws are inserted having a quality level of 2% defectives.
    If the machine cycle time is 3 s, and the down time caused by one defective part is 30 s, determine:
    (a) The percentage downtime on the machine
    (b) The average production rate in assemblies per minute
    If the total machine rate is $16/h (including operator costs and overheads), would it pay to spend $0.20 per 100 screws to improve their quality level to 1% defectives?

31. A three-station free-transfer mechanized assembly machine has a cycle time of 4 s. At each station the ratio of defective to

acceptable parts is 1:150 and the time taken to correct a fault at any station is 10 s. Assuming that all stoppages are due to faulty parts, determine the production rate of the machine for equal buffer spaces between each station for 0, 1, 2, 3, 4, and 5 work carriers.

If the total cost M of running the machine for 1 min is given by

$$M = 0.8 + 0.009b$$

where b is the buffer space between each station, determine the assembly cost per assembly for each of the conditions specified above and hence show that an optimum value of b exists for which the assembly costs are a minimum.

32. An assembly operation involves 10 operations. No operations can be carried out until operation 1 is complete. Operation 10 cannot be carried out until all other operations are completed. Operation 9 cannot be carried out until operations 7 and 8 are complete. Operations 5, 6, and 7 cannot be carried out until operation 3 has been completed. Operation 8 cannot be carried out until operation 4 is complete.

Draw a precedence diagram for this assembly process.

33. Construct a precedence diagram for the 12 parts of the box and lid assembly shown in Fig. P.33. Assume that the first operation is to place the box (part 1) in the work fixture.

34. A feasibility study is to be made for the automatic assembly of three components (Fig. P.34) on an indexing machine that would operate at one cycle per second with an average downtime due to stoppages of 30 s. Preliminary studies indicate that the screw has a quality level of 1.5% completely defective. The plastic base has 2.0% defective but only 0.5% would prevent assembly of the screw; the remainder are cracked or have unacceptable appearance. The metal clips have 1.0% defectives where the screw cannot pass through the hole.

The estimated total rate for the assembly machine is $23/h, including one operator and overhead. It is possible to improve the quality levels of the screw and clip to 0.5 and 0.2%, respectively, by means of automatic inspection machines. How much would you be prepared to spend on these machines to reduce stoppages on the assembly machine? Neglect the cost of dealing with faulty assemblies produced by the machine and assume a payback period of 4000 h and 100% overhead for equipment.

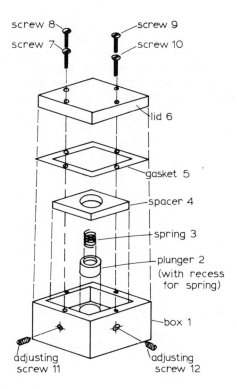

screw 8
screw 7
screw 9
screw 10
lid 6
gasket 5
spacer 4
spring 3
plunger 2
(with recess
for spring)
box 1
adjusting
screw 11
adjusting
screw 12

Fig. P. 33

Fig. P. 34

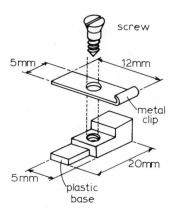

screw
5mm
12mm
metal
clip
20mm
5mm
plastic
base

35.. Given the function of a product and the general appearance required, what kind of considerations should be given to the part and product design to make it suitable for automatic assembly. Irrespective of the ease of automation of the assembly process, only certain types of product are suitable for automatic assembly. List the other necessary features of a product.

Appendix I

Simple Method for the Determination of the Coefficient of Dynamic Friction

Every student of physics or engineering is aware of the difficulties of obtaining an accurate measurement of the coefficient of dynamic friction between two surfaces. The simple methods normally employed in laboratory demonstrations, and with which everyone is familiar, are usually unsatisfactory and yield results with large random errors. The improvement in accuracy that can be obtained through repeated measurement is not usually feasible due to the time-consuming adjustment of the inclination of a plane or the adjustment of the load applied to a slider.

An experimental device that will drive a loaded slider across a surface and record the frictional force resulting is necessarily rather elaborate and expensive, and generally only justifiable when research into the frictional behavior of sliding surfaces is undertaken.

The method described here* is not intended for this kind of work but is suggested as a replacement for simple undergraduate laboratory experiments on friction or for a rapid measurement of the coefficient of dynamic friction, which is so often desirable in any engineering problem involving relative motion between surfaces.

I.1 The Method

Figure I.1 shows the experimental setup where the slider (1) and the straightedge (2) are placed on a sheet of paper on a horizontal and reasonably flat surface such as a drafting board. A means is provided

---

*G. Boothroyd, "Simple Method for the Determination of the Coefficient of Sliding Friction," *Bull. Mech. Eng. Educ.*, vol. 9, 1970, p. 219.

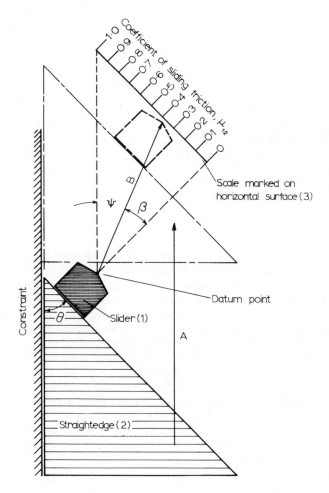

Fig. I.1 Apparatus for determination of coefficient of dynamic friction. (Adapted from G. Boothroyd, "Simple Method for the Determination of the Coefficient of Sliding Friction," *Bull. Mech. Eng. Educ.*, vol. 9, 1970, p. 219.)

for constraining the straightedge to move in a straight line over the surface in a direction inclined at an angle θ to the edge itself. This can readily be arranged using a T-square and triangle.

As the straightedge is moved at a reasonably uniform speed in the direction of arrow A in Fig. I.1, the slider moves in the direction of arrow B, a direction that (as will be shown later) depends only on the coefficient of sliding friction between the slider and the straightedge and the angle θ. Thus, with the appropriate scale and datum point marked on the horizontal surface, the slider may be moved from the datum point to the scale and the coefficient of friction read off directly.

It will be appreciated that with this method, repeated measurements can be made rapidly and a precise determination of the required coefficient of friction thus can be obtained.

The scale marked on the horizontal surface must be appropriate to the chosen angle θ between the straightedge and the direction of its motion. For conditions where the coefficient of friction lies between zero and unity, a convenient scale is obtained when θ is 45 degrees (Fig. I.1). For coefficients of friction greater than unity, a smaller angle of θ must be employed.

I.2 Analysis

The horizontal forces acting on the slider (1) as the straightedge (2) is moved relative to the surface (3) are shown in Fig. I.2. The frictional force between the slider and the surface acts in opposition to

Fig. I.2 Horizontal forces acting on slider.

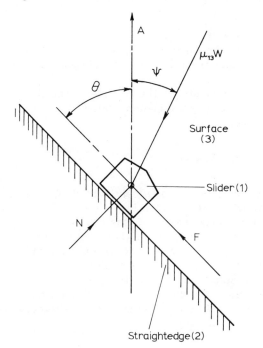

the direction of motion of the slider across the surface. This direction is inclined at an angle ψ to the direction of motion of the straightedge; thus, from Fig. I.2:

$$N = \mu_{13}W \sin (\theta + \psi) \tag{I.1}$$

$$F = \mu_{13}W \cos (\theta + \psi) \tag{I.2}$$

where μ_{13} is the coefficient of sliding friction between the slider and the surface and W is the weight of the slider.
   Now

$$\mu_{12} = \frac{F}{N} \tag{I.3}$$

and hence substitution of Eqs. (I.1) and (I.2) in Eq. (I.3) gives

$$\mu_{12} = \cot (\theta + \psi) \tag{I.4}$$

[note that $(\theta + \psi)$ is the complement of the friction angle β in Fig. I.1]
or

$$\tan \psi = \frac{1 - \mu_{12} \tan \theta}{\mu_{12} + \tan \theta} \tag{I.5}$$

It can now be seen that the direction in which the slider moves is independent of both its weight and the coefficient of friction between the slider and the surface.
   Equation (I.5) may be used to construct the required scale for a given value of θ. If θ is 45 degrees, a convenient value for θ when $0 \le \mu_{12} \le 1.0$, Eq. (I.5) becomes

$$\tan \psi = \frac{1 - \mu_{12}}{1 + \mu_{12}} \tag{I.6}$$

Further, if the scale is arranged parallel to the straightedge, then it is linear with respect to coefficient of friction (Fig. I.1).
   A simpler way of describing the operation of this device is to argue that the slider can move only in the direction of the resultant force applied by the straightedge. This resultant force lies at an angle β (the angle of friction between the slider and the straightedge) to a line drawn normal to the straightedge, and thus the slider moves as shown in Fig. I.1. Further, since the distance read off the scale is proportional to tan β, this scale will be a linear one, giving the coefficient of friction μ_{12}.

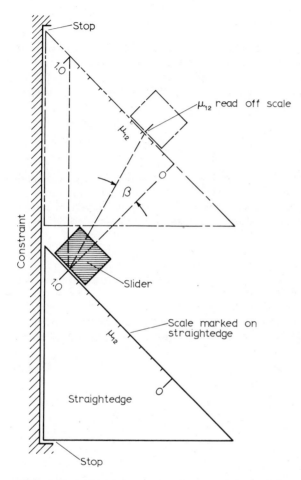

Fig. I.3 Alternative apparatus. (Adapted from G. Boothroyd, "Simple Method for the Determination of the Coefficient of Sliding Friction," *Bull. Mech. Eng. Educ.*, vol. 9, 1970, p. 219.)

An interesting variation of this method is illustrated in Fig. I.3. Here a linear scale in μ_{12} is marked off on the straightedge. The slider is initially set a 1.0 on the scale and the straightedge moved a distance equal to the length of the scale multiplied by $\sqrt{2}$. The final position of the slider will give the value of the coefficient of friction.

I.3 Precision of the Method

Using a celluloid straightedge and a brass slider, 20 measurements of the coefficient of sliding friction were made. It was found that the mean of the readings was 0.207 with 95% confidence limits of ±0.0043 for the mean. This result indicates the relatively high precision that can be obtained.

I.4 Discussion

It is felt that the simple method for the determination of the coefficient of sliding friction between two surfaces described in this appendix holds several advantages over the methods usually employed in under-graduate laboratory demonstrations. These advantages include the following:

1. It requires the minimum of equipment.
2. No delicate adjustments of plane inclination or loading are required.
3. Relatively high precision can be obtained.
4. Repeated readings can be made quickly.
5. Readings are not affected by the static coefficient of friction.
6. Readings are not affected by small variations in the speed of sliding.
7. It is a direct-reading method.

Appendix II

Out-of-Phase
Vibratory Conveyors

Experimental and theoretical investigations* have shown that certain
fundamental limitations exist in the performance of conventional vibratory feeders:

1. The conveying velocity of parts up the inclined track is always
less than that of parts traveling around the flat bowl base. This
means that motion of parts on the track is normally obtained through
the pushing action of those circulating around the bowl bottom. With
this situation there is a tendency for parts to jam in the various selecting and orienting devices fitted to the bowl track.

Some parts, because of their shape, are difficult to feed under
these circumstances. For example, very thin sheet parts are not able
to push each other up the track. In this case the feed rate obtained
with a conventional bowl feeder is very low.

Sometimes it is necessary, as part of the orienting system, to have
a discontinuity in the track. Again because the parts traveling around
the bottom of the bowl cannot push those on the track beyond the discontinuity, the feed rate is generally unsatisfactory.

2. The conveying velocity of parts in a conventional vibratory
bowl feeder is very sensitive to changes in the coefficient of friction
between the part and the track, and conveying velocities are very low
with low coefficients of friction.

3. For high feed rates, it is necessary for the parallel velocity of
the track to be high. However, because of the method of driving a
conventional vibratory bowl feeder, an increase in the amplitude of

---

*A. H. Redford, "Vibratory Conveyors," Ph.D. thesis, Royal College
of Advanced Technology, Salford, England, 1966.

the parallel component of vibration must be accompanied by a corresponding increase in the amplitude of the normal component of vibration. This latter increase is undesirable because as the normal track acceleration increases above the value that causes the component to hop along the track, the mode of conveying quickly becomes erratic and unstable due to the bouncing of parts on impact with the track.

A new drive is now described which is suitable for all types of vibratory conveyors and solves many of the problems associated with conventional designs.

II.1 Out-of-Phase Conveying

The new method for driving vibratory feeding devices is based on the idea that the normal and parallel components of motion of the track should have independent amplitude control and should be out of phase. Under these circumstances the locus of a point on the track becomes elliptical instead of linear.

Fig. II.1 Effect of coefficient of friction in out-of-phase conveying. γ is the phase angle, v_m the mean conveying velocity (in./s), μ the effective coefficient of friction, track angle θ, 4 degrees; vibration frequency f, 21.5 Hz; amplitude ratio a_n/a_p, 0.365; maximum normal track acceleration A_n, 1.2g.

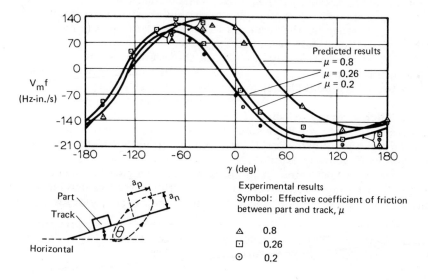

Experimental results
Symbol: Effective coefficient of friction between part and track, μ

△	0.8
▢	0.26
⊙	0.2

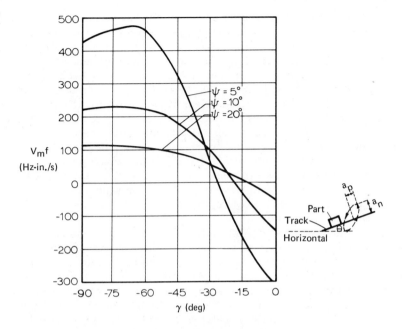

Fig. II.2 Predicted effect of amplitude ratio ($a_n/a_p = \tan \psi$) in out-of-phase conveying; θ = 4 degrees, μ = 0.2, A_n = 1.2$\underline{g}$.

Theoretical and experimental work has been conducted on this new type of drive* and some of its advantages can be demonstrated by means of the results shown in Fig. II.1. In the figure, the product of the mean conveying velocity v_m and the frequency of vibration f is plotted against the phase difference γ between the two components of motion. In the results illustrated, the ratio of the normal a_n and the parallel a_p amplitudes of vibration and the normal track acceleration were both kept constant. The relationships are plotted for three values of the coefficient of friction μ between the part and the track which cover the range likely to be met in practice. It can be seen from the figure that when the phase angle was zero, simulating a conventional feeder, the conveying velocity was very sensitive to changes in μ. Further, for values of μ less than 0.3, the part was moving backward. The results show that if the track parallel motion leads the track normal motion by 65 degrees, the conveying velocity becomes uniformly high for all the values of considered.

*Redford, "Vibratory Conveyors."

Figure II.2 shows the predicted effect on the mean conveying velocity v_m of changing γ in the relevant range (-90 degrees to 0 degrees) for three values of the amplitude ratio a_n/a_p and when the normal track acceleration A_n was kept constant. In these results a track angle of 4 degrees and a coefficient of friction of 0.2 were chosen because it was considered that these represented the most severe conditions likely to be encountered in practice. It is clear from the figure that, for conventional conveying ($\gamma = 0$), as a_p is increased, indicating an increase in the maximum parallel track velocity and a decrease in ψ, the backward conveying velocity of the parts increases. However, for the optimum phase angle ($\gamma = -65$ degrees) the forward conveying velocity is increased as a result of an increase in a_p. It is also of interest to note that if the vibration frequency was 25 Hz (for which a control system has been designed), a conveying velocity as high as 18 in./s (400 mm/s) can be achieved.

These results show that definite advantages are to be gained from operating a vibratory bowl feeder under the optimum out-of-phase conditions. First, the high conveying velocities attainable are almost independent of the nature of the parts being conveyed. Second, because the feed rate can be controlled by adjusting the parallel component of vibration only, the track normal acceleration may be held constant at a level which does not cause erratic movement of the parts (in the results presented the normal track acceleration was 1.2g, which represents stable conveying for most materials). Third, if a_p were to be gradually increased as the part climbs the track, the conveying velocity of the part would gradually increase. This would result in separation of the parts as they climb the track. This situation can be achieved in practice by gradually increasing the track radius and would result in more efficient orienting and greater reliability in operation.

II.2 Practical Applications

Figure II.3 shows an exploded view of a vibratory bowl feeder drive unit designed to operate on the principle outlined above. In this design, motion normal to the track is imparted to the bowl through an intermediate plate supported on the base. Motion parallel to the track is obtained through the springs that support the bowl on the plate. With a suitable controller, the two independent motions will have the required phase difference and the situation described for out-of-phase conveying can be obtained.

Tests conducted on a bowl feeder based on the design proposed above have verified the findings of the research. As a practical example of the capabilities of the new design of feeder, an attempt was made to feed thin mica specimens which a manufacturer had previously

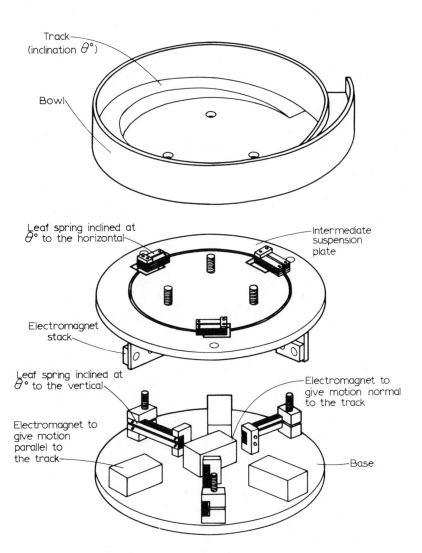

Track
(inclination $\theta°$)

Bowl

Leaf spring inclined at
$\theta°$ to the horizontal

Intermediate
suspension
plate

Electromagnet
stack

Leaf spring inclined at
$\theta°$ to the vertical

Electromagnet to
give motion normal
to the track

Electromagnet to
give motion
parallel to
the track

Base

Fig. II. 3 Exploded view of a vibratory bowl feeder drive which has
independent motion normal and parallel to the bowl track.

found almost impossible to feed in a typical conventional vibratory bowl feeder. With the new feeder, however, it was possible to feed these specimens separately up the track at conveying velocities of up to 18 in./s (400 mm/s) without any erratic motion.

The new type of drive, suitable for all types of vibratory conveyors and called an out-of-phase drive, has many practical advantages. With this type of drive feeders are quieter, greater flexibility in performance can be achieved, more reliable yet more sophisticated orienting devices can be employed, and much higher feed rates can be obtained than with the conventional drive system. (Note: The out-of-phase drive for vibratory conveyors has been patented by the National Research Development Corporation and manufactured by the University of Salford Industrial Centre, Salford, Lancs., England, to whom inquiries should be directed.)

Appendix III

U. Mass. Coding System for the Automatic Feeding and Orienting of Small Parts

III.1 Introduction

The three-digit coding system* presented at the end of this appendix is designed for small parts that are to be automatically fed and oriented. In this system, the basic shape, important features, and various symmetries of a part are described. To use the system it is necessary to understand the meaning of the following terms.

III.2 Terminology

Difficult-to-Feed Parts. In most cases, the feeding and orienting characteristics of a part will be based on its geometry. However, under some circumstances, other features (such as fragility and size) override geometric considerations and can make a part difficult to feed. These parts are given a first digit of 9 and can be described by one or more of the following: flexible, delicate, sticky, light, overlap, large, very small, nest or severely nest, tangle or severely tangle, and abrasive.

For parts with a first digit of 9, there is a separate choice for the second and third digits which defines the appropriate characteristics. When a part is defined as difficult to feed, this does not necessarily mean that feeding and orienting at acceptable rates is impossible but that a highly specialized feeding device is likely to be required.

---

*G. Boothroyd, C. Poli, and L. E. Murch, "The Handbook of Feeding and Orienting Techniques for Small Parts," Department of Mechanical Engineering, University of Massachusetts, Amherst, Mass., 1977.

325

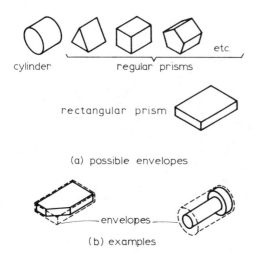

cylinder   regular prisms   etc.

rectangular prism

(a) possible envelopes

envelopes

(b) examples

Fig. III.1 Envelopes.

Envelope. The envelope of a part is the smallest cylinder, regular prism, or rectangular prism that can completely enclose the part (Fig. III.1). When the choice of envelope is not clear, the one that has the maximum contact with the part should be chosen.

Degenerated Envelope. To determine the first digit of the coding system (except for the first digit of 9) it is necessary to know the basic shape of the part and the dimensions of the envelope. To determine the basic shape of a part (a rotational, triangular, and square prismatic or rectangular) it is necessary to examine its degenerated envelope. Essentially, the degenerated envelope is the cylinder, regular prism, or rectangular prism obtained when small projections on the part are neglected. In the example shown in Fig. III.2 the features outside the degenerated envelope are quite small. However, when the features in question are large, a special technique must be

Fig. III.2 Degenerated envelope.

envelope   degenerated envelope

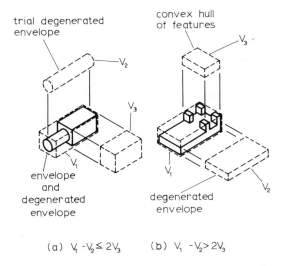

trial degenerated
envelope

V_2

convex hull
of features

V_3

V_3

V_1

envelope
and
degenerated
envelope

V_1

V_2

degenerated
envelope

(a) $V_1 - V_2 \leq 2V_3$ (b) $V_1 - V_2 > 2V_3$

Fig. III. 3 Technique needed to establish the degenerated envelope.

used to establish the degenerated envelope. For example, Fig. III.3a
shows a part that is similar to the part in Fig. III.2 but whose degenerated envelope is the same as the envelope. The cylinder is an unacceptable degenerated envelope because the difference in volume ($V_1 - V_2$)
between the envelope (volume V_1) and the small trial degenerated
envelope (volume V_2) is less than twice the volume V_3 enclosed by the
portion of the part outside the trial degenerated envelope. Thus,
the trial generated envelope is rejected. An example where the envelope
and degenerated envelope are different is shown in Fig. III.3b. It
will be noted in this last example that the volume V_3 is not the actual
volume of the features outside the degenerated envelope. To be precise
it is the volume enclosed by the convex hull of the features and their
projection onto the degenerated envelope, excluding the portion enclosed
by the degenerated envelope (Fig. III.3b). The convex hull of an
object or objects can be imagined as the shape of an elastic film wrapped
around the object or objects. Examples are shown in Fig. III.4.

  With complicated parts the envelope is reduced to a degenerated
envelope by a series of trial inward moves of the envelope surface,
one by one, always keeping the principal axes of the trial degenerated
envelope parallel to the principal axes of the envelope. The first
trial reduction should be made with the surface having the largest
area and where the ratio $V_3/(V_1 - V_2)$ is a minimum (must be less than
0.5). The technique in obtaining the degenerated envelope cannot be
applied simultaneously to two different surfaces. Further examples
are shown in Fig. III.5.

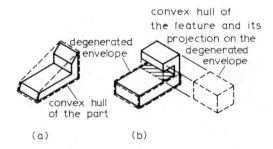

Fig. III.4 Convex hull.

Fig. III.5 Degenerated envelope for various parts.

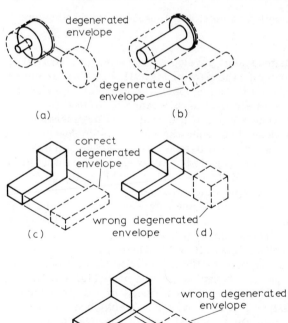

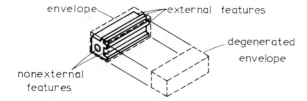

Fig. III.6 Geometric features.

Features. A portion of space or a portion of the part can be called a geometric feature of a part if it satisfies one of the following conditions:

1. The space between the envelope and the degenerated envelope that is not occupied by the part is a feature (for example, a step of a chamfer).
2. The space that is enclosed by the degenerated envelope but is not occupied by the part is a feature (for example, a hole or a groove).
3. The space that is occupied by the part but is not enclosed by the degenerated envelope is a feature (for example, a projection).

In other words, only the space that is both occupied by the part and enclosed by its degenerated envelope is not a geometric feature of the part. A geometric feature that is outside the degenerated envelope is called an external feature; otherwise, it is called a nonexternal feature. A part may have many features which are separated by boundaries of its envelope, degenerated envelope, and itself. An example is shown in Fig. III.6.

Other features, such as paint, lettering, or geometric features that cannot be described macroscopically, such as surface finish, are considered nongeometric features. Except for features that do not require orientation, such as the slot in the head of a screw, all other features that can be used to define the orientation of a part are candidate features for coding. However, only the set of features that can be utilized most efficiently is coded. The importance of a feature is evaluated by the probability that the feature will interact with an orienting device. The less the probability, the less the importance of the feature will be. Therefore, the geometric shape, the location, and the size of a feature are major factors to be considered during the coding process. Some examples are shown in Fig. III.7.

Symmetry. Rotational symmetry is the only geometric symmetry important in this classification of parts. Rotational symmetry means that the orientation of a part is repeated when the part is rotated

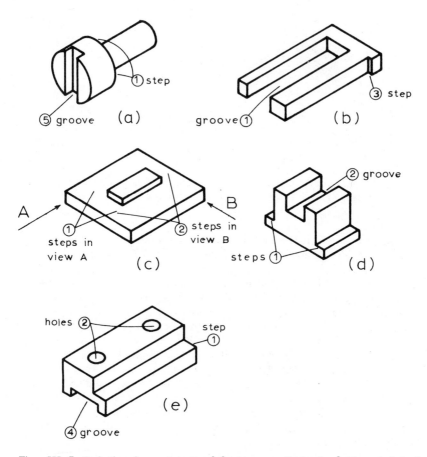

Fig. III.7 Relative importance of features. Key: 1, feature(s) to be
coded; 2, less important feature(s); 3, feature too small; 4, feature
that cannot define the orientation; 5, feature that does not require
orientation and is not coded.

through an angle θ about a certain axis passing through the centroid
of its degenerated envelope. In general, rotational symmetry is not
present unless $2\pi/\theta$ is an integer. Therefore, only those angles equal
to $2\pi/n$ (n is an integer) can be used to examine the rotational symmetry
of a part. Since all parts have 2π (360°) rotational symmetry about
all axes, only those rotational symmetries where $2\pi/\theta$ is an integer
greater than 1 are of interest. There are special cases, such as a
sphere or cylinder, which have an infinitesimally small rotational
symmetry and therefore their orientation is preserved after a rotation

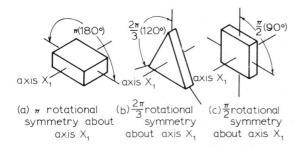

(a) π rotational symmetry about axis X_1 (b) $\frac{2\pi}{3}$ rotational symmetry about axis X_1 (c) $\frac{\pi}{2}$ rotational symmetry about axis X_1

Fig. III.8 Examples of rotational symmetries.

through any angle. A part that has no rotational symmetry except 2π (360°) rotational symmetry is considered as having no rotational symmetry about the axis considered. Figure III.8 shows several examples.

In general, a part may have rotational symmetry about several axes. However, these symmetries are not mutually independent. For instance, if a part has $\pi/2$ (90°) rotational symmetry about two of the three mutually perpendicular axes (Fig. III.9a), it also has $\pi/2$ (90°) rotational symmetry about the third axis. This is also true for π (180°) rotational symmetry (Fig. III.9b). Because of this interdependence of symmetries, the description of the rotational symmetry of a part is considerably simplified. This is based on a subject usually called group theory. An introduction to this theory can be found in any good text on modern algebra.

The rotational symmetry of a feature or features must be defined in relation to an axis passing through the centroid of the degenerated envelope. If a group of features has a certain rotational symmetry, it can be considered as one feature (Fig. III.10). By this convention, at most two features are needed to completely define the orientation of a part.

The features that do not require orientation should be ignored during the coding of the rotational symmetry of a part. Figure III.7a

Fig. III.9 Symmetric axes.

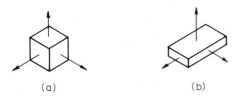

(a)   (b)

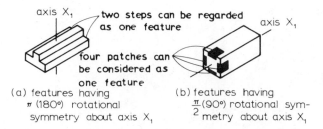

(a) features having
π (180°) rotational
symmetry about axis X₁

(b) features having
$\frac{\pi}{2}$ (90°) rotational symmetry about axis X₁

Fig. III.10 Groups of features having rotational symmetry.

shows an example. The slot in the head of the screw should normally
be disregarded when the rotational symmetry of this particular part
is examined.

For parts with cylindrical or regular prismatic degenerated envelopes,
the axis that is perpendicular to the circular or regular polygonal
cross section of the degenerated envelope and passing through its
centroid is called the principal axis (for example, the X_1 axis in Fig.
III.8b and c). The axes perpendicular to the principal axis passing
through the centroid of the degenerated envelope are called transverse
axes (for example, all axes shown in Fig. III.8b and c except the X_1
axis). If a part has π (180°) rotational symmetry about at least one
of its transverse axes, it does not require orientation end-to-end and
is alpha symmetric. An example is shown in Fig. III.11a. If a part
does not require orientation about its principal axis, it is beta symmetric (Fig. III.11b).

Fig. III.11 Examples of alpha and beta symmetries.

(a) an alpha symmetric part

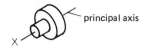

(b) a beta symmetric part

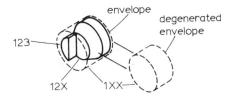

Fig. III.12 Rotational part.

III.3 Examples

In these examples the automatic handling code is determined using the U.Mass. coding system presented at the end of this appendix.

Example 1

Digit 1. The part shown in Fig. III.12 has a cylindrical degenerated envelope and thus the first digit is 0, 1, or 2. The length-to-diameter ratio of the envelope is between 0.8 and 1.5. Thus, the first digit is 1.

Digit 2. The table for the second and third digits for rotational parts is now used. The instructions and notes for the second digit are read starting with 0 until an appropriate digit is found. Since the part in Fig. III.12 has a beta symmetric chamfer on its external surface and it does not satisfy the descriptions for a second digit of 0 or 1, the second digit is 2.

Digit 3. The instructions and notes for the third digit are read starting with 0 until an appropriate digit is found. In the present example, since the projection on the end surface is the main feature that causes beta asymmetry, the third digit is 3.

The code for the part is 123.

Fig. III.13 Square prismatic part.

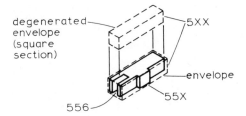

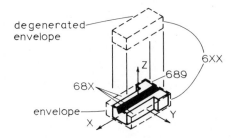

Fig. III.14 Rectangular prismatic part.

Example 2

Digit 1. The part shown in Fig. III.13 has a square prismatic degenerated envelope and thus the first digit is 3, 4, or 5. The length-to-diameter ratio of the smallest cylinder that can completely enclose the part is greater than 1.5. Thus, the first digit is 5.

Digit 2. Since this part has no rotational symmetry about the principal axis which is due to the steps external to the degenerated envelope, the second digit is 5.

Digit 3. Since the through groove on the end surface is the feature that causes alpha asymmetry, the third digit is 6.

The code for the part is 556.

Example 3

Digit 1. The part shown in Fig. III.14 has a rectangular prismatic degenerated envelope and thus the first digit is 6, 7, or 8. Because the ratio A/B is less than 3 and the ratio A/C is greater than 4, the first digit is 6.

Digit 2. Two features are needed to completely define the orientation of this part. One of them is rotationally symmetric about the Z axis and the other (pair) is rotationally symmetric about the X axis. Thus, the second digit is 8.

Fig. III.15 Difficult-to-feed part.

Digit 3. The nongeometric feature that is rotationally symmetric about the Z axis will give the largest third digit, 9.

The code for the part is 689.

Example 4

Since the part shown in Fig. III.15 will severely nest, the code is 9XX. The last two digits are chosen by the special code for parts that are difficult to feed.

The code for the part is 908.

III.4 Coding System for Small Parts for Automatic Handling
     (Choice of the First Digit)

Notes

1. Digits 0 to 8 are reserved for parts that can be fed (but not necessarily oriented) using conventional vibratory or non-vibratory hopper feeders. These digits do not include parts that are flexible, delicate, sticky, light, overlap, large, very small, nest, severely nest, tangle, severely tangle, or are abrasive (see definitions below). Parts having one or more of the foregoing characteristics, irrespective of basic shape, are assigned a first digit of 9.
2. A part whose basic shape (degenerated envelope) is a cylinder or regular prism whose cross section is a regular polygon of five or more sides is called a rotational part.
3. A part whose basic shape (degenerated envelope) is a regular prism whose cross section is a regular polygon of three or four sides is called a triangular or square prismatic part.
4. A part whose basic shape (degenerated envelope) is a rectangular prism is called a rectangular prismatic part.
5. L is the length and D is the diameter of the smallest cylinder that can completely enclose the part.
6. A is the length of the longest side, C is the length of the shortest side, and B is the length of the intermediate side of the smallest rectangular prism that can completely enclose the part.

Definitions

Flexible. A part is considered flexible if the part, or a section of the part, cannot maintain its shape under the action of automatic feeding so that orienting devices cannot function satisfactorily.

Delicate. A part is considered delicate if damage may occur during handling either due to breakage caused by parts falling from orienting sections or tracks onto the hopper base or due to wear caused by recirculation of parts in the hopper. When wear is the criterion, a part would be considered delicate if it could not recirculate in the hopper for 30 min and maintain the required tolerances.

Sticky. If a force, comparable to the weight of a nontangling or nonnesting part, is required to separate it from bulk, the part is considered sticky.

Light. A part is considered too light to be handled by conventional hopper feeders if the ratio of its weight to the volume of its envelope is less than 1.5 kN/m^3.

Overlap. Parts are said to overlap when an alignment of better than 0.2 mm is required to prevent shingling or overlapping during feeding in single file on a horizontal track.

Large. A part is considered to be too large to be readily handled by conventional hopper feeders when its smallest dimension is greater than 50 mm or if its maximum dimension is greater than 150 mm. A part is considered to be too large to be handled by a *particular* vibratory hopper feeder if

$$D + \left(\frac{L}{d}\right)^2 \geq 1.4w$$

where D is the maximum diameter or width of the part measured perpendicular to the feeding direction and in a horizontal plane, L the length of the part measured parallel to the feeding direction, w the width of the track, and d the feeder or bowl diameter.

Very Small. A part is considered to be too small to be readily handled by conventional hopper feeders when its largest dimension is less than 3 mm. A part is considered to be too small to be readily handled by a *particular* vibratory hopper feeder if its largest dimension is less than the radius of the curved surface joining the hopper wall and the track surface measured in a plane perpendicular to the feeding direction.

Nest. Parts are considered to nest if they interconnect when in bulk, causing orientation problems. No force is required to separate the parts when they are nested.

Severely Nest. Parts are considered to severely nest if they interconnect and lock when in bulk and require a force to separate them.

Tangle. Parts are said to tangle if a reorientation is required to separate them when in bulk.

Severely Tangle. Parts are said to severely tangle if they require manipulation to specific orientations and a force is required to separate them.

Abrasive. A part is considered to be abrasive if it may cause damage to the surfaces of the hopper feeding device unless these surfaces are specially treated.

III.4.1 First Digit

parts can easily be fed (but not necessarily oriented) using conventional hopper feeders (see note 1)	rotational parts (see note 2)	L/D < 0.8 disks (see note 5)	0
		$0.8 \leq L/D \leq 1.5$ short cylinders (see note 5)	1
		L/D > 1.5 long cylinders (see note 5)	2
	triangular or square prismatic parts (see note 3)	L/D < 0.8 flat parts (see note 5)	3
		$0.8 \leq L/D \leq 1.5$ cubic parts (see note 5)	4
		L/D > 1.5 long parts (see note 5)	5
	rectangular parts (see note 4)	$A/B \leq 3$, A/C > 4 flat parts (see note 6)	6
		A/B > 3 long parts (see note 6)	7
		$A/B \leq 3$, $A/C \leq 4$ cubic parts (see note 6)	8
	parts are difficult to feed using conventional hopper feeders (see note 1)		9

III.4.2 Parts with First Digit of 0, 1, or 2: Rotational Parts
(See Note 1)

Notes

1. A rotational part is one whose basic shape (degenerated envelope)
   is a cylinder or a regular prism having five or more sides. In
   addition, the part is not difficult to feed.
2. The part does not require orientation end to end.
3. A main feature causing alpha asymmetry is one that defines the
   end-to-end orientation of the part.
4. These are parts that will orient themselves with their principal
   axis vertical when placed in a parallel sided horizontal slot.
5. A beta symmetric step or chamfer is a concentric reduction in
   diameter. The cross section can be circular or any regular polygon
   of four or more sides. Discrete projections, recesses or irrelevant
   features, should be ignored in choosing this digit.
6. The reductions and increases in diameter forming the groove must
   be concentric. The cross sections can be circular or any regular
   polygon of four or more sides. Discrete projections, recesses,
   or irrelevant features should be ignored in choosing this digit.
7. Basically, these parts have an alpha symmetric external shape,
   but their center of mass is not at the geometric center of the part.
8. If exposed features are prominent but the asymmetry caused by
   these features is too small to be employed for orienting purposes,
   the asymmetry is said to be slight asymmetry.
9. A beta symmetric part does not require orientation about its
   principal axis.
10. A main feature causing beta asymmetry is one that completely
    defines the orientation of the part about its principal axis.
11. Some parts can only be fed one way. However, when a choice
    exists, the technique employed and hence the code can be affected
    by the delivery orientation.

III.4.3 Parts with First Digit of 0, 1, or 2: Second Digit

part is alpha symmetric (see note 2)		0
[code the main feature or features, causing alpha asymmetry] part is not alpha symmetric (see note 3)	Part can be fed in a slot supported by large end or protruding flange with center of mass below supporting surfaces (see note 4)	1
	beta symmetric steps or chamfers on external surfaces (see note 5)	2
	beta symmetric grooves, holes, or recesses (see note 6) — on both side and end surfaces	3
	beta symmetric grooves, holes, or recesses (see note 6) — on side surface only	4
	beta symmetric grooves, holes, or recesses (see note 6) — on end surfaces only	5
	beta symmetric hidden features with no corresponding exposed features (see note 7)	6
	beta asymmetric features or beta symmetric parts with features other than steps, chamfers, or tapers but too small for orientation purposes (see note 9)	7
	beta symmetric step, chamfer, or taper too small for orientation purposes	8
	other features, slight asymmetry, features too small or nongeometric features [such as paint, lettering, etc.] (see note 8)	9

III.4.4 Parts with First Digit of 0, 1, or 2: Third Digit

part is beta symmetric (see note 9)	to be fed end to end (see note 11)			0
	to be fed side by side (see note 11)			1
[code the main feature or features causing beta asymmetry] part is not beta symmetric (see note 10)	beta asymmetric projections (can be seen in silhouette)	on side surface only		2
		on end surfaces only		3
		on both side and end surfaces		4
	beta asymmetric grooves, holes, recesses on external surfaces	through grooves can be seen in end view		5
		through grooves can be seen in a side view	on end surfaces	6
			on side surface	7
		holes or recesses [cannot be seen in outer shape of silhouette in end views]		8
	other features, slight asymmetry, features too small, or nongeometric features [such as paint, lettering, etc.]			9

III.4.5 Parts with First Digit of 3, 4, or 5: Triangular and
Square Prismatic Parts (See Note 1)

Notes

1. A part whose basic shape (degenerated envelope) is a regular
prism whose cross section is an equilateral triangle or square is
called a triangular or square prismatic part. However, parts that
are difficult to feed are excluded.
2. Part does not require orientation about its principal axis.
3. A part has rotational symmetry about a specified axis if the part's
orientation is repeated by rotating it through a certain angle
(less than 360°) about that axis. Examples are as follows:

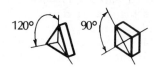

4. When the degenerated envelope of a part is a perfect cube, the
principal axis should be selected according to the following
priorities:
(1) any axis about which the part has 90° rotational symmetry;
(2) an axis about which the part has 180° rotational symmetry and
clearly not 90° rotational symmetry; (3) an axis about which the
part has 180° rotational symmetry and almost 90° rotational symmetry; (4) when a part has no rotational symmetry and there is
more than one main feature, the principal axis should be the axis
of symmetry of one of the main features. When utilizing these
rules and multiple choices still exist, the axis that will provide
a code with the smallest third digit should be selected as the
principal axis
5. Part does not require orientation end to end (it has 180° rotational
symmetry about at least one transverse axis).
6. A main feature causing alpha asymmetry is one that defines the
end-to-end orientation of the part and distinguishes the end and
side surfaces.
7. The various aspects of a part resting on a plane are called natural
resting aspects.
8. If exposed features are prominent but the asymmetry caused by
these features is too small to be employed for orienting purposes,
the asymmetry is said to be slight asymmetry. For the part that
is 180° rotationally symmetric about a certain axis, slight asymmetry

implies that the part is almost 90° rotationally symmetric about the same axis.

9. Steps, chamfers, or through grooves are features that result in a deviation of the silhouette of the part from the silhouette of its envelope.

10. These are parts that will orient themselves with their principal axis vertical when placed in a parallel-sided horizontal slot.

11. For parts with through grooves that can be seen in both side and end views, the grooves that can be seen in the end view should be coded if the first digit is 3. Otherwise, the grooves that can be seen in the side view should be coded.

12. A feature is said to be too small if that feature is too small to be employed for orientation purposes.

13. The same orientation of the part will be repeated only once by rotating the part through 180° about its principal axis.

14. The same orientation of the part will not be repeated by rotating the part through any angle less than 360° about its principal axis.

15. For parts with features on both side and end surfaces, the feature or features on the end surfaces should be coded if the first digit is 3. Otherwise, the feature or features on the side surfaces should be coded.

III.4.6 Parts with First Digit of 3, 4, or 5: Second Digit

	part has 90° or 120° rotational symmetry about the principal axis (see notes 2, 3, and 4)		0
part has 180° rotational symmetry about the principal axis [code the main feature or features causing 180° rather than 90° rotational symmetry about the principal axis] (see notes 4 and 13)	steps or chamfers can be seen in side or end views (see note 9)		1
	through grooves can be seen in side or end views (see note 9)		2
	holes or recesses [cannot be seen in outer shape of silhouette]		3
	other features, features too small, or slight asymmetry (see notes 8 and 12)		4
part does not have 180° rotational symmetry about the principal axis [code the main feature or features causing rotational asymmetry] (see notes 4 and 14)	steps or chamfers can be seen in side or end views (see note 9)	external to the degenerated envelope	5
		nonexternal	6
	through groves can be seen in side or end views (see note 9)		7
	holes or recesses [cannot be seen in outer shape of silhouette]		8
	other features, features too small or slight asymmetry (see notes 8 and 12)		9

III.4.7   Parts with First Digit of 3, 4, or 5: Third Digit
If Second Digit Is 0 (For notes, see Section III.4.5)

part is alpha symmetric (see note 5)		part has only one natural resting aspect or end and side surfaces can be readily distinguished by their shapes or dimensions (see note 7)	0
		end and side surfaces can be distinguished because of steps, chamfers, holes, or recesses	1
		end and side surfaces can only be distinguished because of other features, features too small or slight asymmetry (see notes 8 and 12)	2
[code the main feature or features causing alpha asymmetry] part is not alpha symmetric (see note 6)	steps or chamfers (see note 9)	part can be fed in slot and supported by large end or protruding flanges with center of mass below supporting surfaces and the part is not triangular (see note 10)	3
		part cannot be fed in slot and supported by large end or protruding flanges with center of mass below supporting surfaces or part is triangular	4
	through grooves (see note 9)	can be seen in end view (see note 11)	5
		can be seen in side view (see note 11)	6
	holes or recesses (cannot be seen in outer shape or silhouette)		7
	other geometric features		8
	slight asymmetry, features too small or non-geometric features (such as paint, lettering, etc.) (see notes 8 and 12)		9

III.4.8 Parts with First Digit of 3, 4, or 5: Third Digit
If Second Digit Is 1-9 (For notes, see Section III.4.5)

part is alpha symmetric (see note 5)	code the same feature or features coded in the second digit	steps, chamfers, or grooves can be seen in side view or other features on side surfaces (see note 9)	0
		steps, chamfers, or grooves can be seen in end view or other features on end surfaces (see note 9)	1
	end and side surfaces can only be distinguished because of features too small or slight asymmetry (see notes 8 and 12)		2
part is not alpha symmetric [code the main feature or features causing alpha asymmetry] (see note 6)	steps or chamfers provided by external features (see note 9)	features on side surfaces (see note 15)	3
		features on end surfaces (see note 15)	4
	steps or chamfers provided by nonexternal features (see note 9)	features on side surfaces (see note 15)	5
		features on end surfaces (see note 15)	6
	holes or recesses [cannot be seen in outer shape of silhouette]	on side surfaces (see note 15)	7
		on end surfaces (see note 15)	8
	other features, slight asymmetry, or features too small (see notes 8 and 12)		9

III.4.9 Parts with First Digit of 6, 7, or 8: Rectangular
Prismatic Parts (See Note 1)

Notes

1. A part whose basic shape (degenerated envelope) is a rectangular
prism is called a rectangular part. However, parts that are
difficult to feed are excluded.
2. 180° rotational symmetry about an axis means that the same orientation of the part will be repeated only once by rotating the part
through 180° about that axis.
3. Part can be oriented without utilizing features other than the
dimensions of the envelope.
4. Steps, chamfers or through grooves are features that result in a
deviation of the silhouette of a part from the silhouette of its
envelope.
5. If exposed features are prominent but the asymmetry caused by
these features is too small to be employed for orientation purposes,
the asymmetry is said to be slight asymmetry. For a part that
has 180° rotational symmetry about a certain axis, slight asymmetry
implies that the part has almost 90° rotational symmetry about that
axis.
6. A feature is said to be too small if it is too small to be employed
for orientation purposes.
7. A part having no rotational symmetry means that the same orientation
tion of the part will not be repeated by rotating the part through
any angle less than 360° about any one of the three axes X, Y,
and Z.
8. A main feature is a feature that is chosen to define the orientation
of the part. All the features that are chosen to completely define
the orientation of the part should be necessary and sufficient for
the purpose. Often, features arise in pairs or groups and the
pair or group of features is symmetric about one of the three axes
X, Y, and Z. In this case, the pair or group of features should
be regarded as one feature. Using this convention, two main
features at most are needed to completely define the orientation
of a part.
9. Sometimes, when a part has no rotational symmetry, its orientation
can either be defined by one or by two main features. Under these
circumstances the part code is determined by the following in de-
creasing order of preference: (1) choose one main feature if it
results in a third digit less than 5; (2) choose two main features
if they result in a third digit less than 5; (3) choose one main
feature if it results in a third digit greater than 5; (4) choose two
main features if they result in a third digit greater than 5.
10. The symmetric plane is the plane that divides the part into halves
which are mirror images of each other.

III.4.10 Parts with First Digit of 6, 7, or 8: Second Digit

part has 180° rotational symmetry about all three axes (see note 2)			0
part has 180° rotational symmetry about one axis only (see note 2)	about X axis		1
	about Y axis		2
	about Z axis		3
part has no rotational symmetry [code the main feature or features that can completely define the orientation] (see notes 7 and 8)	part's orientation is defined by one main feature only (see note 9)	part has a symmetric plane (see note 10)	4
		part has no symmetric plane (see note 10)	5
	part's orientation is defined by two main features and at least one of them is a step, chamfer, or through groove, or a group of such features (see note 9)	one feature is symmetric about X axis and the other one is symmetric about Y axis	6
		one feature is symmetric about Y axis and the other one is symmetric about Z axis	7
		one feature is symmetric about Z axis and the other one is symmetric about X axis	8
part has slight asymmetry about at least one of its axes or the orientation of the part can only be defined by two main features, neither of which are steps, chamfers, or through grooves (see notes 5 and 8)			9

III.4.11 Parts with First Digit of 6, 7, or 8: Third Digit
If Second Digit Is 0 (For notes, see Section III.4.9)

two or more adjacent surfaces of the envelope have similar dimensions (code the main feature or features which distinguish the adjacent surfaces having similar dimensions)	three adjacent surfaces of the envelope have significant differences in dimensions (see note 3)		0
	steps or chamfers (see note 4)	parallel to X axis	1
		parallel to Y axis	2
		parallel to Z axis	3
	through grooves (see note 4)	parallel to X axis	4
		parallel to Y axis	5
		parallel to Z axis	6
	holes or recesses (cannot be seen in outer shape of silhouette)		7
	slight asymmetry or features too small (see notes 5 and 6)		8
	other geometric features or nongeometric features (such as paint, lettering, etc.)		9

III.4.12 Parts with First Digit of 6, 7, or 8: Third Digit
If Second Digit Is 1-9 (For notes, see Section III.4.9)

[code the feature that gives the largest third digit, if more than one feature is utilized to define the orientation of the part] (see note 8) code the main feature			
	steps or chamfers (see note 4)	parallel to X axis	0
		parallel to Y axis	1
		parallel to Z axis	2
	through grooves (see note 4)	parallel to X axis	3
		parallel to Y axis	4
		parallel to Z axis	5
	holes or recesses [cannot be seen in outer shape of silhouette]		6
	other geometric features		7
	features too small (see note 6)		8
	nongeometric features [such as paint, lettering, etc.]		9

III.4.13 Parts with First Digit of 9: Difficult-to-Feed Parts;
Second Digit

			nonflexible	0
parts are small and nonabrasive	parts do not tend to overlap during feeding	not delicate		
			flexible	1
		delicate	nonflexible	2
			flexible	3
	parts tend to overlap during feeding	not delicate	nonflexible	4
			flexible	5
		delicate	nonflexible	6
			flexible	7
parts are very small or large but are nonabrasive				8
abrasive parts				9

III.4.14 Parts with First Digit of 9: Third Digit If
Second Digit Is 0 through 7

		not sticky	0
parts will not tangle or nest	not light	sticky	1
	light	not sticky	2
		sticky	3
parts will tangle or nest but not severely	not light	not sticky	4
		sticky	5
	light	not sticky	6
		sticky	7
parts will severely nest but not severely tangle			8
parts will severely tangle			9

III.4.15 Parts with First Digit of 9: Third Digit If
Second Digit Is 8

very small parts	rotational parts	disks or short cylinders L/D ≤ 1.5	0
		long cylinders L/D > 1.5	1
	nonrotational parts	flat parts A/B ≤ 3 A/C > 4	2
		long parts A/B > 3	3
		cubic parts A/B ≤ 3 A/C ≤ 4	4
large parts	rotational parts	disks or short cylinders L/D ≤ 1.5	
		long cylinders L/D > 1.5	6
	nonrotational parts	flat parts A/B ≤ 3 A/C > 4	7
		long parts A/B > 3	8
		cubic parts A/B ≤ 3 A/C ≤ 4	9

III.4.16 Parts with First Digit of 9: Third Digit If Second
         Digit Is 9

parts will not severely tangle or nest	small parts	orientation is defined by geometric feature(s) alone	nonflexible	parts do not tend to over-lap during feeding	0
				parts tend to overlap during feeding	1
				flexible	2
		orientation is not defined by geo-metric feature(s) alone		parts do not tend to over-lap during feeding	3
				parts tend to overlap during feeding	4
	large parts			part's orientation is defined by geometric feature(s) alone	5
				part's orientation is not defined by geo-metric feature(s) alone	6
	very small parts			part's orientation is defined by geometric feature(s) alone	7
				part's orientation is not defined by geo-metric feature(s) alone	8
parts will severely tangle or nest					9

Appendix IV

Laboratory Experiments

This appendix gives a complete description of two typical laboratory experiments that may be included in a college or university course on automatic assembly. The first experiment is designed to illustrate certain practical aspects of the performance of a vibratory bowl feeder; it also indicates how the results of the tests are best presented to gain the maximum information. Clearly, similar experiments could be designed to study the performance of other types of parts feeding devices employed in automatic assembly. The second experiment illustrates how the coefficient of dynamic friction between small parts and a feed track may be obtained. This information is used in the second part of the experiment to verify the predictions of the theoretical analysis of a horizontal delivery gravity feed track presented in Chapter 6.

IV.1 Performance of a Vibratory Bowl Feeder

IV.1.1 Object

To determine (1) the relationship between vibration amplitude and feed rate for a constant bowl load and (2) the effect of bowl loading on the performance of a vibratory bowl feeder.

IV.1.2 Equipment

Vibratory bowl feeder (10-in. or 12-in. bowl); 1000 mild-steel parts of 5/16 in. diameter and 1 in. long; transducer arranged to measure the vertical component of the bowl vibration amplitude; stopwatch.

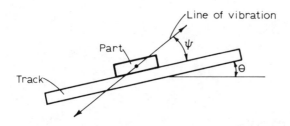

Fig. IV.1 Section of track in vibratory bowl feeder. θ is the track angle, ψ the vibration angle.

IV.1.3 Procedure

1. For a range of settings on the bowl amplitude control and with a bowl load of 500 parts, measurements are made of the time taken for a part to travel between two marks scribed on the inside of the bowl. The vertical vibration amplitude of the bowl is also measured, and in these tests a device is fitted to the top of the bowl track which continuously returns the parts to the bottom of the bowl in order to maintain the bowl load constant.
2. Commencing with the bowl full (1000 parts) and the amplitude control set to give a low feed rate (less than 1 part per second), the times are measured for successive batches of 100 parts to be delivered. In this test the parts are allowed to pass down the delivery chute and the readings are continued until the bowl becomes empty.

IV.1.4 Theory

Theoretical and experimental work has shown that the parameters which affect the mean conveying velocity v_m in vibratory conveying are:

1. Maximum track acceleration A (m/s^2)
2. Operating frequency ω (rad/s)
3. Track angle θ (see Fig. IV.1)
4. Vibration angle ψ (see Fig. IV.1)
5. Coefficient of friction between component and track μ
6. Acceleration due to gravity g (m/s^2)

Dimensional analysis may now be applied to this problem as follows. Let

$$v_m{}^{a_1} = f[A^{a_2}, \omega^{a_3}, \theta^{a_4}, \psi^{a_5}, \mu^{a_6}, g^{a_7}] \qquad \text{(IV.1)}$$

Using the fundamental dimensions of length (L) and time (T), Eq. (IV.1) becomes

$$\left(\frac{L}{T}\right)^{a_1} = f\left[\left(\frac{L}{T^2}\right)^{a_2}, \left(\frac{1}{T}\right)^{a_3}, \left(\frac{L}{T^2}\right)^{a_7}\right] \tag{IV.2}$$

Since the terms θ, ψ, and μ are dimensionless, they have been omitted here to simplify the work. Thus, for Eq. (IV.1) to be dimensionally homogeneous,

$$a_1 = a_2 + a_7 \quad \text{and} \quad -a_1 = -2a_2 - a_3 - 2a_7 \tag{IV.3}$$

or

$$a_7 = a_1 - a_2 \quad \text{and} \quad a_3 = -a_1 \tag{IV.4}$$

Substituting Eqs. (IV.4) into Eq. (IV.1) yields

$$v_m^{a_1} = f[A^{a_2}, \omega^{-a_1}, \theta^{a_4}, \psi^{a_5}, \mu^{a_6}, g^{a_1-a_2}]$$

or rearranging terms with similar exponents, we obtain

$$\left(\frac{v_m \omega}{g}\right)^{a_1} = f\left[\left(\frac{A}{g}\right)^{a_2}, \theta^{a_4}, \psi^{a_5}, \mu^{a_6}\right] \tag{IV.5}$$

Thus, for a given bowl and given parts, the dimensionless conveying velocity $v_m\omega/g$ is a function of the dimensionless maximum track acceleration A/g. The theoretical work described in Chapter 3 shows that it is more convenient to employ the component A_n of acceleration normal to the track. However, it was also shown that conveying is generally achieved by the pushing action of the parts circulating around the flat bowl base. In this case, the effective track angle is zero and therefore A, in the analysis above, should be taken as the vertical component A_v of the bowl acceleration.

If it is assumed that the bowl moves with simple harmonic motion, the maximum bowl acceleration may be obtained from measurements of the vertical bowl amplitude and a knowledge of the operating frequency ω.

In the second part of the experiment, the mean feed rate F_0 for each increment in bowl load will be given by

$$F_0 = 100/t_f \quad \text{parts/s} \tag{IV.6}$$

where t_f is the time taken (in seconds) to feed 100 parts.

IV.1.5 Presentation of Results

Figures IV.2 and IV.3 show results obtained with a typical commercial bowl feeder. In Fig. IV.2, the dimensionless mean conveying velocity $v_m\omega/g$ is plotted against the dimensionless vertical bowl acceleration A_V/g. It can be seen that feeding occurs for all values of A_V/g greater than 0.32, and from this it is possible to estimate the coefficient of static friction μ_s between the parts and the track using the analysis described in Chapter 3. Thus,

$$\mu_s = \frac{\cot \psi}{g/A_0 - 1} \tag{IV.7}$$

where ψ is the vibration angle, A_0 the minimum vertical acceleration of the bowl for feeding to occur, and g the acceleration due to gravity. In the results presented here, μ_s was estimated to be 0.95 for a mild-steel part in a rubber-coated bowl.

Figure IV.3 shows the changes in Feed rate F_0 as the bowl gradually empties. It can be seen that as the bowl load reduced, the feed rate increased rapidly. Clearly, when the bowl is empty, the feed rate will have fallen to zero. For the amplitude setting employed in this test, the bowl could be used to feed the cylindrical parts to a machine

Fig. IV.2 Effect of vertical bowl acceleration on conveying velocity for a commercial vibratory bowl feeder. Mild-steel cylindrical parts 5/16 in. dia. × 1 in. long; bowl load 500; spring angle 65 degrees; vibration frequency 50 Hz.

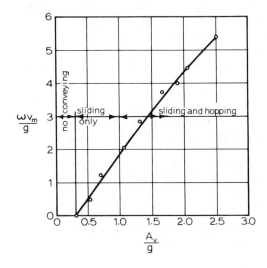

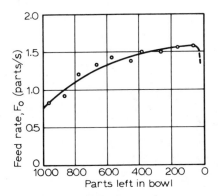

Fig. IV.3 Load sensitivity of commercial vibratory bowl feeder. Mild-steel cylindrical parts 5/16 in. dia. × 1 in. long; rubber coated track; spring angle 65 degrees; vibration frequency 50 Hz.

or workhead requiring about 40 parts per minute. It is clear that, because of the large increase in feed rate as the bowl empties, excessive recirculation of the parts would occur.

IV.2 Performance of a Horizontal Delivery Gravity Feed Track

IV.2.1 Objectives

(a) To determine the coefficient of dynamic friction μ_d between the parts and feed track used in the experiment
(b) To examine experimentally and theoretically the performance of a horizontal delivery gravity feed track

IV.2.2 Equipment (a)

The equipment used in the determination of the coefficient of dynamic friction consisted (Fig. IV.4) of a straight track whose angle of inclination θ to the horizontal could be varied between 20 and 60 degrees. The equipment is designed to record accurately the time taken for a part to slide from rest a given distance down the track. Near the top of the track, a short peg projecting upward through a hole in the track retains the part until the spring cantilever supporting the peg is deflected by depressing the button. The deflection of the cantilever also closes a pair of contacts at the moment the part is released; this initiates the count on a digital clock. A further contact is positioned at the bottom of the track a distance L from the front of

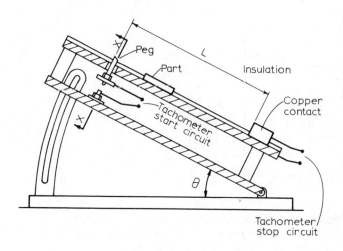

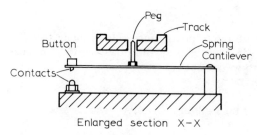

Enlarged section X–X

Fig. IV.4 Apparatus used in the determination of the coefficient of dynamic friction.

the retaining peg. When the part arrives at the bottom of the track it completes the circuit, which stops the count on the clock.

IV.2.3 Theory (a)

The equation of motion for a part sliding down an inclined track is

$$m_p a = m_p g \sin \theta - \mu_d m_p g \cos \theta$$

or

$$\frac{a}{g} = \sin \theta - \mu_d \cos \theta \qquad \qquad (IV.8)$$

where m_p is the mass of the part, θ the track inclination, a the acceleration of the part, and μ_d the coefficient of dynamic friction between the part and the track.

For a straight inclined track the acceleration of the part is uniform and the time t_s taken for the part to slide a distance L is given by

$$t_s^2 = \frac{2L}{a} \tag{IV.9}$$

Combining Eqs. (IV.8) and (IV.9) gives

$$\mu_d = \frac{\sin\theta - 2L/gt_s^2}{\cos\theta} \tag{IV.10}$$

IV.2.4 Procedure (a)

In the present experiment, where the part was of mild steel and the track of aluminum alloy, the times were measured for the part to slide a distance of 4 in. (101.6 mm) with the track angle set at 30, 45, and 60 degrees. For each condition, 20 readings were taken and averaged and the 95% confidence limits for each average were computed using the t statistics. The corresponding values of μ_d were computed using Eq. (IV.10).

IV.2.5 Results (a)

The mean values of μ_d obtained for each track angle are presented in Table IV.1 together with their corresponding 95% confidence limits. Since no significant variation of μ_d with changes in track angle was evident, it was thought reasonable to take the average of all the readings obtained. This yielded a mean value of μ_d of 0.353 with a range, for 95% confidence, of 0.325 to 0.381.

IV.2.6 Equipment (b)

Figure IV.5 shows the design of the experimental horizontal delivery feed track. A parts release and timing arrangement similar to that used in the first part of the experiment is provided. In this case the column of parts is retained by a peg positioned on the horizontal section 2 in. (50.8 mm) from the beginning of the curved section, and the

Table IV.1 Mean Values of μ_d and 95% Confidence Limits

θ (degrees)	μ_d	95% confidence limits
30	0.358	±0.028
45	0.354	±0.012
60	0.347	±0.006

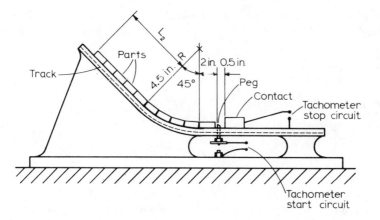

Fig. IV.5 Apparatus used to investigate the performance of a gravity feed track.

contact that arrests the column of parts and stops the count is positioned 0.5 in. (12.7 mm) from the front of the peg.

IV.2.7 Theory (b)

The derivation of the theoretical expression for the initial acceleration on release of a column of parts held in a feed track of the present design was developed in Chapter 6. Thus,

$$
\frac{a}{g} = \left(L_2(\sin \alpha - \mu_d \cos \alpha) - L_1\mu_d e^{\mu_d\alpha} \right.
$$
$$
+ \frac{R}{1 + \mu_d^2} \left[(1 - \mu_d^2)(e^{\mu_d\alpha} - \cos \alpha) - 2\mu d \sin \alpha \right] \right)\left(L_2 \right.
$$
$$
\left. + L_1 e^{\mu_d\alpha} + \frac{R}{\mu_d} (e^{\mu_d\alpha} - 1) \right)^{-1} \qquad (IV.11)
$$

Substitution of the values of L_1 = 2 in. (50.8 mm), R = 4.5 in. (114.3 mm), and α = 45 degrees for the experimental rig and the mean, upper, and lower values of μ_d = 0.353, 0.325, and 0.381, respectively, obtained in the first part of the experiment gives the required predicted relationship between a/g and L_2.

In the experiment the time t_p(s) was recorded for the column of parts to slide a distance of 0.5 in. (12.7 mm). Since this distance was small compared with the dimensions of the feed track, it could be

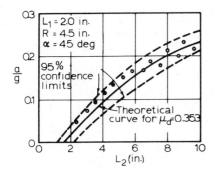

Fig. IV.6 Performance of a horizontal delivery gravity feed track (steel parts on aluminum track).

assumed that the acceleration a of the column of parts was constant. Thus,

$$\frac{a}{g} = \frac{2(0.5)}{gt_p^2} = \frac{2.59 \times 10^{-3}}{t_p^2} \tag{IV.12}$$

IV.2.8 Procedure (b)

The time was recorded for the parts to slide 0.5 in. (12.7 mm) for a range of values of L_2. For each condition an average of 20 readings was obtained.

IV.2.9 Results (b)

The experimental results are plotted in Fig. IV.6 together with the two curves representing the 95% confidence limits. It can be seen that all the experimental results fall within the two theoretical curves representing the 95% confidence limits for the mean value of μ_d.

IV.2.10 Conclusions

1. The coefficient of friction between the parts and the track used in the experiment has been successfully determined.
2. The experimental results obtained for the acceleration of a column of parts in a horizontal delivery gravity feed track fall within the range, for 95% confidence, of predicted values using the results of the analysis and the values of μ_d obtained in the first part of the experiment. This confirms that the theory, determined for parts of an infinitesimally small length, is valid for parts of finite length provided that this length is small compared with the dimensions of the track.

3. The performance of parts in a horizontal delivery gravity feed track can be accurately estimated using the theoretical equation, provided that the coefficient of dynamic friction between the parts and the track is known.

Nomenclature

A	length of longest side of rectangular envelope; maximum track acceleration
A_i	basic cost of one part
A_n	component of maximum track acceleration normal to the track of a vibratory feeder
A_p	component of maximum track acceleration parallel to the track of a vibratory bowl feeder
B	length of intermediate side of rectangular envelope; rate of increase in cost of one part due to quality level
C	length of shortest side of rectangular envelope; capital cost of equipment, including overhead
C_B	cost of transfer device per workstation or buffer space for a free-transfer machine
C_c	cost of work carrier
C_{dA}	cost of programmable robot or workhead with d degrees of freedom
C_F	cost of automatic feeding device and delivery track
C_g	cost of special gripper per part to be handled or cost of gripper to handle one part only
C_i	cost of one part
C_M	cost of manually loaded magazine

C_{PF} cost of programmable feeder

C_{pr} average production (assembly) cost for one assembly

C_s cost of workstation for single-station operator assembly

C_{sg} cost of special gripper to handle all parts in an assembly

C_T cost of transfer device per workstation for an indexing machine

C_t total cost of each assembly produced, including the cost of parts

$C_{t(min)}$ minimum total cost of the completed assembly

C_{ug} cost of a universal gripper

C_W cost of dedicated workhead

D diameter of part; track depth; proportion of downtime

D_h diameter of rivet head

E efficiency of feeder; modified efficiency of feeding system

E_{ab} energy barrier in moving a part from orientation a to orientation b

E_{ba} energy barrier in moving a part from orientation b to orientation a

F frictional resistance or force; feed rate

F' unrestricted feed rate

F_m minimum unrestricted feed rate (that is, the unrestricted feed rate when bowl is full)

F_{max} maximum feed rate

FR_s feed rate per slot for an external gate hopper

H height the part hops on the track

J effective distance the part hops on the track

J_0 J/R

L length of part

L_p number of parts used to refill a bowl feeder

L_1 length of horizontal track section

L_2 length of straight inclined track section

M number of parts in a full bowl feeder; cost of operating a machine per unit time if only acceptable assemblies are produced

M_t	total cost of operating a machine per unit time, including operator's wages, overhead, actual operating costs, machine depreciation, and cost of dealing with unacceptable assemblies
N	normal force; number of vanes; number or parts in hopper; number of assemblies
N_a	number of parts in aspect a
N_b	number of parts in aspect b; number of parts below lower sensor and acceptable level
N_d	number of product design changes during the life of an assembly machine
N_p	number of parts delivered per blade, per cycle, per slot or per revolution; number of parts held in feed track; number of different products to be assembled during the life of an assembly machine
P	number of parts fed from bowl feeder during time t; number of parts per vane falling on empty rail
P_a	production rate of acceptable assemblies; probability for aspect a; probability for orientation a
P_b	probability for aspect b; probability for orientation b
P_u	number of unacceptable assemblies produced per unit time
Q	equivalent cost of one operator on one shift in terms of capital equipment
R	partition ratio $P_a/(P_a + P_b)$; radius of gravity feed track; radius of base of truncated cone; radius of track
R'	average recirculation per part in a bowl feeder
R_e	probability that a part will be rejected
R_L	load ratio for stationary hook hopper
S	bowl load sensitivity; number of shifts
T	machine downtime due to one defective part
T_c	time taken by operator to dismantle an unacceptable assembly
V_a	annual production volume
W	weight of part, width of part; assembly operator's rate, including overhead
W_t	total rate for all operators engaged on an assembly machine
a	linear acceleration of part; average number of assemblies present in a buffer storage on a free-transfer machine

a_n normal component of amplitude of track vibration

a_0 amplitude of track vibration

a_p parallel component of amplitude of track vibration

a_s center distance between slots of external gate hopper

b   distance from the apex of the vee cutout orienting device to the bowl wall; size of buffer storage between workheads on a free-transfer machine

b_0 b/R

b_t track width

b_u largest value of b_0 for which all the unwanted orientations of a truncated cone are rejected

b_w smallest value of b_0 for which all the wanted orientations of a truncated cone are accepted

c   clearance

d   hopper diameter; barrel diameter; downtime on machine due to faulty parts when N assemblies are produced; number of degrees of freedom of programmable workhead (or robot); diameter of screw head

d_t minimum track diameter

f   frequency of vibration

g   acceleration due to gravity

g_n component of g normal to the track

h   depth of screw head

h_g gap between cylinder and sleeve in external gate hopper

i   number of stations on individual indexing machines connected by buffer storage (hybrid machine)

i_{opt} optimum number of stations for each indexing unit on a hybrid machine

j   distance as defined in Fig. 5.22

j_0 j/R

k   number of parts assembled by each operator on an assembly line or number of parts to be assembled by individual assembly robots

ℓ length of slot; length of track; width of vanes; distance from the screw head bottom to the center of mass

$\bar{\ell}$	average part length
ℓ_b	length of barrel
m	proportion of defective parts causing a machine fault
m_p	mass of part
m_1	part mass per unit length
n	reciprocation frequency; rotational frequency; number of parts to be assembled in one assembly
n_c	critical rotational frequency
n_f	average number of parts fed during one workhead cycle with a 100% efficient feeder
n_M	number of parts contained in one magazine
n_{max}	maximum reciprocation frequency; maximum rotational frequency
n_s	number of standard deviations
p	number of parts fed in time t; proportion of part presentation stations where a programmable feeder is used; $[2 + (y/x)^2]^{1/2}$
q	$[1 + (y/x)^2]^{1/2}$; proportion of parts for which a gripper change is necessary
r	radius from center of hopper; blade radius; radius of the top of a truncated cone; reliability of part delivery; radial position
r_b	centerboard hopper swing radius
r_h	radius of hopper hub
r_0	r/R
s	slot width; shank diameter
t	time; cycle time of assembly workhead; average assembly time per part assembled for a programmable workhead or robot; diameter of top of scrw head
t_f	total period of feeder cycle
t_g	time for a robot to change grippers
t_i	time for index
t_ℓ	time taken by operator to load one part into a magazine, on-line
t_ℓ'	time taken by operator to load one part into a magazine, off-line

t_0 average time taken by operator to assemble one part if no mechanical assistance is provided

t_0' average time taken by operator to assemble one part if parts are presented in correct orientation

t_p time for a part to move one part length

t_{pr} average production (assembly) time for one acceptable assembly

t_s time taken for parts to slide from slot of rotary disk hopper feeder

t_w workhead cycle time

t_1 time to lift blade of centerboard hopper

t_2 dwell time for blade of centerboard hopper

v peripheral velocity of sleeve in external gate hopper; velocity of part; conveying velocity

v_m mean conveying velocity

w width of blade

x half-width of a square prism; ratio of defective to acceptable parts

x_{opt} optimum ratio of defective to acceptable parts giving minimum total cost of the completed assembly

y half-length of a square prism; number of parts available divided by number of parts assembled

α vane angle; inclination angle of gravity feed track; angle between screw axis and a line normal to the track; cot (L/D)

α_1 arcsin (1/q)

α_2 arcsin (1/p)

β arctan (x/y); friction angle

β_w wiper blade jamming angle

γ riser angle; phase angle

η efficiency of bowl feeder orienting system

θ track angle; inclination of delivery chute; inclination of elevator hopper; half-angle of the vee cutout orienting device

θ' inclination of disk in rotary disk feeder

θ_m maximum inclination of track

θ_T	critical track angle where screw just touches track cover (or tilt angle)
θ_w	angle between the wiper blade and the bowl wall
λ	hopper inclination for external gate hopper
μ	effective coefficient of friction
μ_b	coefficient of dynamic friction between part and spinning disk
μ_d	coefficient of dynamic friction
μ_{max}	maximum value of coefficient of static friction for sliding to occur
μ_p	coefficient of dynamic friction between part and moving surface
μ_r	coefficient of dynamic friction between part and hook
μ_s	coefficient of static friction
μ_w	coefficient of dynamic friction between part and hopper wall
ϕ	angle of hopper wall
ϕ_g	gate angle for external gate hopper
ψ	vibration angle
ψ_{opt}	optimum vibration angle
ω	angular frequency of vibration; angular velocity

Index

Active orienting device, 101,
107–108
Air assisted feed track, 183–184
Alpha symmetry, 332
Analysis of orienting systems,
108–117, 137–142
Annual production volume, 226,
245–254, 279–280
Assembly cost per part (*see*
Cost of assembly per as-
sembled part)
Assembly costs using
free transfer machines, 222
hybrid machines, 224
indexing machines, 207–208,
222, 276
operator assembly, 225–234
programmable workheads,
228–234
robots, 234–240
Assembly machines
free transfer, 25, 201, 209–
218, 220, 222–224, 230,
240–254, 291–294
indexing, 13--19, 201–209,
220, 222–224, 230, 240–
254, 288, 291

[Assembly machines]
in-line, 9–13
rotary, 9–13
Assembly methods, 5
Automatic assembly advantages,
5
Average part length, 141

Beta symmetry, 332
Bladed wheel hopper feeder,
86–89
Buffer storage of assemblies,
201, 209, 210, 214–217,
221

Cams
centerboard hopper feeder,
56
crossover, 20, 22–24
Centerboard hopper feeder
blade frequency, 59–60
efficiency, 60–63
feed rate, 60
general features, 56
load sensitivity, 60–61
maximum track inclination,
56–60

Centrifugal hopper feeder, 76–79

Coefficient of dynamic friction, 157, 313–318, 361

Computer simulation of
free transfer systems, 218
ON-OFF controls, 161

Cost of
assembly, 206, 207
assembly machine, 218
assembly per assembled part, 225–227, 241–254
capital equipment, 218
component, 205–207
downtime, 207
feeder, 219, 221, 238, 244
gripper, 229, 235, 244
magazine, 100, 229, 235, 244
operating machine, 206
operator, 206, 218–221, 229–230
parts quality, 206–207
programmable feeder, 237–238, 241
robot, 235, 241, 244
special gripper, 236
transfer device, 221, 244
universal gripper, 237, 241, 244
work carrier, 221, 244
workhead, 221, 244
workstation, 244

Cut-out orienting device
scallop, 104–105, 111
V-shaped, 104–105, 117–126

Cycle time (*see also* Production rate), 202, 210

Defective parts (*see also* Quality level of parts), 201–204, 207–213

Degenerated envelope, 326–329

Degrees of freedom, 194, 229

Design for assembly, 200, 255–262, 266–267

Design for Assembly Handbook, 267, 297

Design of assembly machines, 275–298

Design of components for
assembly, 255–262
feeding and orienting, 262–267

Design of products for assembly, 254

Dimensional analysis, 356–357

Downtime, 202–203, 207–217, 276–279, 288–294

Economic comparisons of assembly machines, 240–254

Economics of indexing machines, 202, 205

Economics of various machines, 294–297

Edge riser, 108, 137–138, 144–145

Effective coefficient of friction, 163–164

Effective hop, 112, 117–118

Efficiency
centerboard hopper feeder, 60–63
external gate hopper feeder, 67–69
rotary disk feeder, 73, 75
centrifugal hopper feeder, 78–79
stationary hook hopper feeder, 86–87

Elevator hopper feeder, 96–98

Endless conveyor belt, 96, 99

Energy barrier, 128–134

Envelope, 326–329

Equivalent cost of capital equipment, 218–244

Escapements
drum, 188–190
drum spider, 188–189
gate, 190
jaw, 190–192
ratchet, 185–186
rotary table, 184–185
slide, 187–188
star-wheel, 188–189
worm, 188–189

Evans, Oliver, 3

External gate hopper feeder
  general features, 65
  feed rate, 65—68
  maximum peripheral velocity,
    65, 71, 72
  efficiency, 67—69
  load sensitivity 68—69
  optimum inclination, 70

Feasibility study, 279—297
Feed tracks (*see also* Gravity
  feed tracks), 100
Feeding
  angled parts, 92
  cylinders, 55, 63, 68, 72,
    75, 77, 79, 86, 88, 91, 97, 98
  disks, 55, 62, 76, 81, 96
  headed parts, 63, 64, 68, 71,
    87
  nuts, 88, 95
  square prisms, 56, 97
  V-shaped parts, 91, 94
Feeding and orienting
  cups, 105, 125
  feasibility study, 285—286
  parts with steps and grooves,
    107
  rectangular blocks, 108
  screws, 102
  truncated cones, 105, 125
  U. Mass. coding system for,
    262, 274, 325—354
  V-shaped parts, 106
  washers, 103—104
Feed rate
  bladed wheel hopper feeder,
    88
  centerboard hopper feeder,
    60
  centrifugal hopper feeder,
    76—79
  elevator hopper feeder, 98
  external gate hopper feeder
    65—72
  reciprocating tube hopper
    feeder, 55
  revolving hook hopper feeder,
    80—81

[Feed rate]
  rotary centerboard hopper
    feeder, 93
  rotary disc hopper feeder,
    73—74
  stationary hook hopper feeder,
    85
  tumbling barrel hopper feeder,
    89—92
Ford, Henry, 3
Free transfer machines, 201,
    209—218, 220, 222—224,
    230, 240—254, 291—294
Frictional effects
  centerboard hopper feeder,
    56—57
  centrifugal hopper feeder, 76
  reciprocating tube hopper
    feeder, 54
  rotary disk feeder, 70, 73
  stationary hook hopper feeder,
    83—85

Gravity feed tracks, 149—180,
    359—364
Gravity feed tracks for headed
    parts 163—180
Geneva mechanism, 70, 93
Gripper design, 200, 234
Group Technology, 194

Handbook of Feeding & Orienting
    Techniques for Small Parts,
    125, 262
Hard surface, 128, 135—136
Hierarchical control, 196—198
Hopper load
  centerboard hopper feeder,
    68—69
  reciprocating tube hopper
    feeder, 54
  rotary disk feeder, 75
  tumbling barrel hopper feeder,
    89
Hybrid system, 223—224, 240—
    254

In-bowl tooling, 101

Indexing machines, 201–209, 220, 222–224, 230, 240–254, 288–291
Indexing mechanisms
  crossover cam, 20, 22–24
  Geneva, 20, 22–23
  rachet and pawl, 20–22
  rack and pinion, 20–21
  requirements, 19–20
Initial distribution matrix, 114–115, 141
Inspection, 52, 100

Jamming
  centerboard hopper feeder, 64
  external gate hopper feeder, 65
  reciprocating tube hopper feeder, 53
  stationary hook hopper feeder, 85

Laboratory experiments, 355–364
Level sensing devices, 102
Load sensitivity
  centerboard hopper feeder, 60–61
  external gate hopper feeder, 68–69
  rotary disk feeder, 75
  vibratory bowl feeder, 41–48

Machine layout, 286–294
Machine pacing, 25
Magazines, 99–100
Magnetic disk feeder, 94–97
Magnetic elevating hopper feeder, 99
Maximum frequency
  centrifugal hopper feeder, 78–79
  magnetic disk feeder, 94
  revolving hook hopper feeder, 81
  rotary hook hopper feeder, 93
Memory pin system, 277
Minimum cost of assembly, 207, 209

Modified system efficiency, 141–142
Modular pick and place, 194
Multistation operator assembly, 225–226, 240–254

Narrow track, 102, 106–108, 137–138, 143, 145
Natural resting aspects, 110, 126–137
Natural resting aspects for
  cylinders, 134, 136
  rectangular prism, 138–139
  solid regular prism, 134–136
  square prism, 123, 136
Nomogram for gravity feed track design, 156
Nonvibratory feeders
  classification, 52
  general principles, 51–52

ON-OFF sensors, 157–162
Operator pacing, 25
Optimum angle for
  centerboard hopper feeder, 56–60
  external gate hopper feeder, 69–70
Optimum indexing unit size, 224, 230
Optimum parts quality, 207–209
Orientations, 109–110, 126–127, 138–141
Orienting device matrix, 114, 138
Orienting system, analysis of, 108–117, 137–142
Orienting system efficiency, 112–113, 115, 141
Orienting system feed rate, 141
Orienting system matrix, 114–115, 138
Out-of-bowl tooling, 101, 145–148
Out-of-phase vibratory conveyor, 319–324

Parts motion in a vibratory bowl feeder, 112, 117–118
Parts placing mechanisms, 191–194
Passive orienting devices, 101
Performance of
  indexing machines, 201–205
  free transfer machines, 210–218
Phase angle, 320–322
Pick and place, 192–194
Position of workheads on free transfer machines, 216–218
Powered feed tracks, 180–183
Precedence diagram, 280–283
Pressure brake, 102–103
Product design changes, 231–232, 245
Product style variations, 232–234, 245
Production time
  free transfer, 221
  hybrid, 223
  indexing machine, 203–206
  operator assembly, 225–228
  with programmable workheads, 228–234
  with robots, 234–240
Programmable assembly
  advantages, 230
  economic comparisons, 241–254
  multistation, 228–234
  single-station, 234–240
PUMA, 195–200
Push and guide, 192

Quality level of parts, 201–208, 210–216, 217, 234–235, 238, 244, 283–285, 288

Reciprocating fork hopper feeder, 64
Reciprocating tube hopper feeder applications, 54–55

[Reciprocating tube hopper feeder]
  feed rate, 55
  general features, 54
Recirculation (nonvibratory feeders), 51
Remote center compliance, 200
Reprogrammable pick and place, 194
Revolving hook hopper feeder, 80–81
Rivets (*see* Feeding headed parts)
Robots, 195–200, 239–254
Root, Elihu, 3
Rotary centerboard hopper feeder, 93–94
Rotary disk feeder
  continuous, 74
  indexing, 70–73

Single-station operator assembly, 227–228, 240–254
Sloped track and ledge, 103, 112
Slot orienting device, 102
Soft surface, 127–135
Spiral elevators, 46
Stationary hook design, 82–85
Stationary hook hopper feeder, 82–86
Step device 111–117
Symmetry, 329–332

Taylor, F. W., 3
Track angle, 27, 32, 35–36, 48
Transfer mechanisms
  chain-driven work carrier, 13, 17, 19
  indexing, 19
  pawl, 15–16
  shunting work carrier, 13, 15, 17, 18
  walking beam, 13–15
Transfer systems
  continuous, 9
  intermittent, 9, 12

Tumbling barrel hopper feeder,
89−92

U. Mass. coding system for
automatic assembly, 267−274
feeding and orienting, 262,
274, 325−354
Unacceptable assemblies, 204,
206
Universal assembly center,
241−254
Unrestricted feed rate (nonvibratory feeder), 51

VAL, 199−200
Vibration angle, 27, 31−32, 34,
37−39
Vibratory bowl feeder
delivery rate, 44
feeder design, 48
general features, 27−48
laboratory experiments, 355−
359
load detectors, 46
natural frequency of vibration,
46−47
ON-OFF controls, 47
recirculation, 43, 45
SCR drives, 47−48
unrestricted feed rate, 36,
42−44

Vibratory conveying
amplitude of vibration, 29−30,
38−41, 48
effective hop, 40−41
frequency of vibration, 27, 32−
34, 38−39, 48
frictional effects, 29, 32, 37,
48
hop, 30−32, 34, 40, 41
mean conveying velocity, 32−
35, 37−40, 48
mechanics, 27−32
optimum vibration angle, 34−35,
48
parts motion, 29−32, 34, 40−
41, 43
track acceleration, 29−37
track angle, 27, 32, 35−36,
48
vibration angle, 27, 31−32,
34, 37−39

Wall projection and narrowed
track, 107
Whitney, Eli, 2
Wiper blade, 102, 103, 108,
137−138, 142, 145
Work carrier, 9
Workheads, 43
Working range, 121−122